6D97
£40-00

D1149215

THE HACCP
Food Safety Manual

T.A.S.C. LIBRARY LEEDS

202690 4

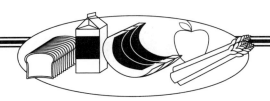

The HACCP
Food Safety Manual

JOAN K. LOKEN, C.F.E.

John Wiley & Sons, Inc.

New York • Chichester • Brisbane • Toronto • Singapore

363.1927 LOK
(BIN) 106624
202690

T A S C
LIBRARY
L R S

This text is printed on acid-free paper.

Copyright © 1995 by Joan K. Loken, C. F. E.

All rights reserved. Published simultaneously in Canada.

Reproduction or translation of any part of this work beyond
that permitted by Section 107 or 108 of the 1976 United
States Copyright Act without the permission of the copyright
owner is unlawful. Requests for permission or further
information should be addressed to the Permissions Department,
John Wiley & Sons, Inc., 605 Third Avenue, New York, NY
10158-0012.

This publication is designed to provide accurate and
authoritative information in regard to the subject
matter covered. It is sold with the understanding that
the publisher is not engaged in rendering legal, accounting,
or other professional services. If legal advice or other
expert assistance is required, the services of a competent
professional person should be sought.

Library of Congress Cataloging-in-Publication Data:
Loken, Joan K.
    The HACCP food safety manual / Joan K. Loken.
        p.   cm.
    Includes bibliographical references.
    ISBN 0-471-05685-5 (pbk. : acid-free)
    1. Food handling—Handbooks, manuals, etc. 2. Food
service—Sanitation—Handbooks, manuals, etc.   I. Title.
TX537.L58   1994
664' .0028'9—dc20   94-17915

Printed in the United States of America

10 9 8 7 6 5 4 3 2

Contents

Preface

A powerful new wave is about to hit the foodservice industry and regulatory agencies — HACCP (Hazard Analysis Critical Control Points), a risk-based focus on food management systems. Using this logical, practical, low-cost, and commonsense approach, we can achieve food safety through anticipation and prevention of hazards.

How can HACCP benefit you? Using it

1. Prevents foodborne illness

2. Makes life easier

3. Improves results of health inspections

4. Enhances guest satisfaction

Times have changed, and we can no longer be complacent about food safety. We live in an environment that presents more hazards and risks than existed 10 years ago. Customers have become more aware of and concerned about the potential dangers of unsafe food and foodborne illness. They want something to be done about this problem. The future for prevention and control is HACCP. Those who are proactive in applying this system in their businesses will be more successful.

To make HACCP work, three things must happen:

1. Focus on the critical hazards in your foodservice operation.

2. Take HACCP down to the simplest level so it is understandable to the food handler.

3. Make HACCP a part of the operation.

General Mills, one of the leaders in HACCP implementation, defines a "critical hazard" as an imminent health hazard or a guest dissatisfaction. For example, if cooling is the issue, focus on cooling. If foreign objects get into food, focus on the causes and prevention. Since incorporating this system totally into its operations, General Mills has increased food quality between 17 and 30 percent This means more satisfied, and thus more frequent, diners. Studies show that if a customer rates an operation as "Excellent" on a comment card, he or she is six to seven times more likely to return than if the rating is "Very Good."

This book has evolved from HACCP training courses I have given over the past four years. The training has been developed and refined in a cooperative effort of industry professionals working with federal, state, and local regulators. As the collection of training materials grew, participants requested that they be made available in book form. Preliminary versions of this manual have been used by

- Industry professionals, in learning how to implement HACCP
- Educators and trainers, in teaching HACCP
- Regulatory agencies, in working with industry

Although the HACCP system is basically simple and straightforward, it is easy to forget some of the steps and procedures in an attempt to apply them at once. Steps 1 through 7 of this book are designed to help you get started by breaking down the concepts into tasks that you can apply immediately. Then, as you refine and develop the system for your own food management system, you can build on them. The HACCP system should be customized to be as practical and compatible as possible with your operation. The eighth section, "Foodborne Illness," explains the relationship between food handling practices and the causes of food-borne illness. When food handlers understand why certain practices may be unsafe, they are better motivated to accept preventive procedures. This section answers questions about food-borne illness and provides illustrative graphics and charts. It also covers crisis management, explaining how to respond to complaints and how to lessen the effects of those complaints, and helps you establish a crisis management plan.

Employee training is essential for a successful food safety program. The ninth section, "Employee Training Materials," will stimulate food safety training ideas for your operation. It contains many ideas that you can incorporate into your own system. A listing of sources for HACCP training, includes videotapes—an excellent vehicle when accompanied by an introduction, discussion of major points, summary, and evaluation. The final sections of this manual include a bibliography, a glossary, and "Food Protection Quizzes."

With this book you may successfully implement a HACCP food management system or an HACCP training program that will be reflected in guest safety, customer satisfaction, and improved employee performance and morale.

Joan K. Loken

Acknowledgments

This manual has been a collaborative effort of many individuals and agencies whose guidance, wisdom, advice, and support have made it possible. My thanks go to:

Sheila Kane, M.P.H., and Raenette Hamann, M.B.A., R.D., Environmental Health Specialists, and Jim Austin, Consumer Protection Chief, The Denver Department of Public Health /Consumer Protection, for the development and continued support of the Managers' Food Safety Certification Program and for editorial support of this manual.

D. J. Inman, Southwest Regional Food Specialist, U.S. Food and Drug Administration, for the planning and development of the food safety seminars and for sharing his presentation slides, which are included in this manual and usable as training overheads.

Chuck Higgins, U.S. Food and Drug Administration State Trainer, who first enlightened me to HACCP, for sharing the most recent FDA writings on HACCP, which are the foundation of this manual.

Patti Klocker, Senior Consumer Health Protection Specialist, Colorado Department of Public Health, and Environment for editorial support in updating material to the newest HACCP thinking.

Joan Hohn, Hospitality Project Manager; Ray Langbehn, Department Chair, Hospitality, Meeting, Travel Administration, Metropolitan State College of Denver; and Wendy Foster, R.D., American Society of Hospital Food Service Administrators Association, for their special editorial assistance.

Rod Fouts, computer guru, who is always there to teach me new techniques.

Above all, I am grateful to my husband, Bruce, for always being supportive and encouraging.

How to Use This Manual

This manual is designed to be your working document for food safety. Use it for ideas, posters, overheads, and training materials. Although you are encouraged to make copies of training aids for your own use, you may not copy the entire manual for distribution.

Organization

This manual is organized in the logical HACCP steps each of which is introduced by a short description, followed by a practical application. There are also supporting discussions, posters, charts, and work book projects to help you better understand the HACCP process.

Training Aids

Included with each step are pages of supporting information in a format that allows them to be used as posters or overheads to assist you in your training program.

Workbook Projects

Also included is a section of Employee Training Materials to be used as training aids and to help you better understand the process.

Foodborne Illness

This section includes training aids, a discussion of foodborne illness, and food microbiology reference material. It also includes a discussion of how to prepare for and handle foodborne illness complaints.

Bibliography

The most recent training aids available, as well as regulatory agency resources, are listed here.

Posters/Overheads

Anticipating Paradigm Shifts

Paradigm: A shared set of assumptions. The paradigm is the way we perceive the world; water to fish. The paradigm explains the world to us and helps us predict its behavior. ADAM SMITH, *Powers of the Mind*

Paradigms. We all develop our own perspectives and unwritten rules—the way we do things—and so do our organizations. The new buzzword describing these rules is "paradigms" (pair-a-dimes). Specifically "paradigms" are rules we take for granted —our most basic assumptions about how we live and work. Paradigms help us to evaluate and organize new information quickly.

Paradigm paralysis. But paradigms can also have a limiting effect. Our paradigms may be so deeply rooted, so unquestioned, that they can become barriers to our ability to see new opportunities.

The paradigm effect. The "old way" of doing something may seem the "only way." We can be blinded to new opportunities and solutions. Whatever field or profession you're in, understanding the influence of this paradigm effect is absolutely vital to the organization's future. JOEL BARKER, *Discovering the Future*

We live in a time of rapid and far-reaching evolution, in which everything from population demographics to technology and tastes are changing at unprecedented speed. In the next decade, the concepts with which we are familiar are going to be changing continually. If managers want to be successful, they will have to learn to anticipate and facilitate change—no matter how fast it comes.

The increased focus on the mechanisms of disease and the preventive factors surrounding them are prompting regulators to incorporate the HACCP approach in their inspections. Sanitation considerations that meet HACCP guidelines will be designed into facilities. All operators recognize that value perception diminishes with poor sanitation standards.*

Old Paradigm	New Paradigm
Sanitation	Food safety
44-point inspection	HACCP, preventing disease
Cover, put in refrigerator	Rapid cool to 40°F, cover, refrigerate
Hair restraints	Hand washing critical
Physical environment	Food handling processes
Temperatures Hot and Cold	Continuous system, total time in danger zone

FDA's Vision for Foods Safety**

In August 1993, Michael R. Taylor, Deputy Commissioner for Policy of the U.S. Food and Drug Administration, defined the FDA's vision for food safety:

> Our current system of food protection is not well designed to prevent food safety problems. When we inspect a facility, we can visually determine with reasonable confidence how it is operating during the time we are there, but we have little basis for knowing what the conditions were prior to the inspection or for predicting with confidence what the conditions will be in the future. . . .
>
> Fortunately, the answer seems clear. And the answer is one whose core conceptual elements we did not have to invent. I am referring to HACCP—a concept that was invented 30 years ago and is already in use by many food companies.
>
> At the heart, HACCP is just the application of good common sense to the production of safe food. Possible avenues of hazard are identified; appropriate preventive controls are designed and installed; and controls are monitored and records are kept to assure that the system is working properly; and, when problems occur, they are identified and promptly corrected.
>
> But, while simple in its basic concepts, HACCP has features that make it a sophisticated and very powerful tool for meeting our food safety responsibility. It is, science based . . . preventive . . . recognizes where the responsibility lies. . . .
>
> We envision a transformation of federal food inspection, based on HACCP, in

---

*Anticipating Paradigm Shifts, *Food Management,* January 1993.
**FDA's Plans for Food Safety and HACCP: Institutionalizing a Philosophy of Prevention.*

THE PARADIGM PARODY

SOME PEOPLE MAKE THINGS HAPPEN

SOME PEOPLE THINK THEY MAKE THINGS HAPPEN

SOME PEOPLE WATCH THINGS HAPPEN

SOME PEOPLE WONDER WHAT HAPPENED

SOME PEOPLE DON'T KNOW ANYTHING HAPPENED AT ALL

which food inspectors no longer rely solely on visual inspections of facilities, laboratory examination of food, and correction of problems after the fact, but instead focus on verifying that well designed systems for preventive controls—HACCP plans—are in place and functioning properly.

Food Safety vs. Sanitation

The management of food safety is changing. Is your system of management the traditional sanitation system, or have you made a step forward to a food safety management system? Test to see whether your food safety management system is up to date. Answer yes or no, regarding implementation of the following items in your operation.

1. Do you pay more attention to the cleanliness of your facility than to food temperatures?

2. Do you rapidly cool foods to 40°F within four to six hours and rapidly reheat foods to 165°F within two hours?

3. Do you have a procedure for recording time and food temperatures at cool down, reheating, and hot holding?

4. Do you know at which critical control points (steps) in your food preparation system you are at highest risk for cross-contamination?

5. Do you check to see whether your refrigerator adequately cools foods to 40°F in four to six hours?

6. Do you have a system to test and check that sanitizers are at proper concentrations and are being properly used?

7. Do you have an enforced policy and procedure to ensure proper hand washing?

8. Do you have an enforced policy to determine when employees are sick or have flu-like symptoms?

9. Have you written control procedures into your recipes for temperature, handling, reheating, and use of leftovers to ensure that food safety standards are known and followed?

10. Are you aware of the changes in our commercial food growing and processing systems that have food safety implications?

If you answered No to

1–2 questions: You are knowledgeable and incorporating most current standards in your food safety management system.

2–4 questions: You are in the process of updating your food safety management system.

5 or more questions: It is highly recommended that you update your current food safety management system.

The new way of thinking is FOOD SAFETY. It is concerned with the integrity of food processing systems, which includes sanitation as part of the systems. However food safety requires a more critical look at the factors known to cause foodborne illness. Five factors are known to cause 80 percent of foodborne illness outbreaks:

- Improper cooling
- Advance preparation (12 hours' lapse)
- Infected persons
- Inadequate reheating
- Improper hot holding

HACCP, Hazard Analysis Critical Control Points, is a food safety system that focuses on potentially hazardous foods—and how they are handled in the food service environment. Self-inspection is the critical ingredient in HACCP. The basic HACCP concepts dovetail with TQM (total quality management) strategies. This system uses a flow chart to identify steps that are likely to cause failure in a process and to develop procedures to lower risks.

In order for a successful HACCP program to be implemented, management must be committed to HACCP. A commitment by management indicates an awareness of the benefits sand costs of HACCP, which includes education and training of employees. Benefits, in addition to food safety, are a better use of resources and timely response to problems.

The focus of an HACCP inspection is on how food is handled, not on how clean the walls and floors are. The result is safer food handling and, consequently, safer food.

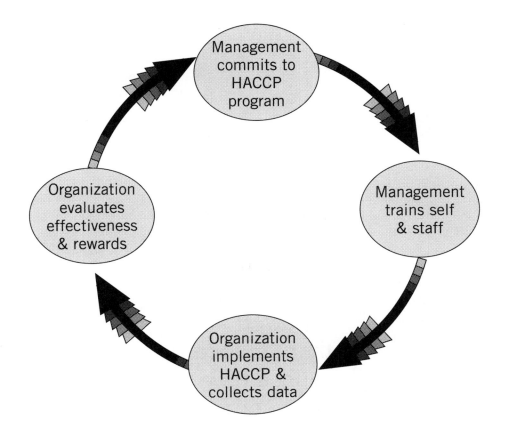

OUR MISSION

To Minimize Consumer Risk of
Illness and Injury from
Foods Consumed in
Food Establishments

*T*here is
nothing more
powerful
than an idea
whose time
has come.

–Victor Hugo

Introduction

What Is HACCP?

HACCP (pronounced "Has-sip") is a difficult name for a simple and effective way to ensure food safety. HACCP stands for the "Hazard Analysis and Critical Control Points" system. It allows you to predict risks to food safety and prevent them before they happen. By using HACCP, you no longer have to rely solely on routine inspections to spot potential food safety hazards.

The HACCP process of ensuring food safety was developed in the 1960s by the Pillsbury Company as part of its effort to produce foods for the space program. (Imagine how serious it would be if astronauts got food poisoning in space.) Pillsbury developed a system to predict and present safety problems throughout the food preparation process.

This system identified potential problems with food safety in advance and set up methods to control each possible hazard. The company kept records to make sure the controls worked. With the HACCP system, Pillsbury, made safe foods. Testing for safety was unnecessary; the HACCP system prevented food safety problems.

Today many food companies use the HACCP system to make sure their products are safe. The U.S. Food and Drug Administration, the U.S. Department of Agriculture, and the U.S. Department of Commerce all encourage HACCP safety plans for food processing in retail food stores, restaurants, and food processing plants.

This system can be used in conjunction with present food protection programs or Total Quality Management (TQM) programs. HACCP takes into consideration

- Factors that contribute to most outbreaks
- Risk assessment techniques to identify and prioritize hazards

Food safety is a key to good business. The sale of unsafe foods can cause waste, illness, lost sales, and lost customers. Keeping foods safe means jobs, good business, happy customers, and greater profitability.

HACCP is, first and foremost, a proactive concept. The technique based on it treats the production of food as a total, continuous system, assuring food safety from harvest to consumption. Included in this system are purchasing, receiving, storage, preparation, and service. Each of these components is evaluated by principles of failure analysis. The premise is simple: If each step of the process is carried out correctly, the end product will be safe food.

What Are Hazard Analysis Critical Control Points (HACCPs)?

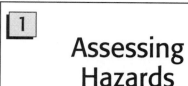 **1** Assessing Hazards

→ **2** Identifying CCPs (critical control points)

↓

 3 Setting Up Procedures for CCPs

↓

 4 Monitoring CCPs

↓

 5 Taking Corrective Action

↓

 6 Setting Up Record-Keeping System

← **7** Verifying the System Is Working

HACCP

IS AN EVALUATION SYSTEM

TO IDENTIFY

TO MONITOR

TO CONTROL

CONTAMINATION RISKS IN FOODSERVICE ESTABLISHMENTS

BASIC HACCP PROCEDURES

1. Identify potentially hazardous foods and sensitive ingredients.

2. Find sources and specific points of contamination.

3. Determine the potential for microorganisms to

   a. survive a heat process and

   b. multiply at room temperature and during hot and cold storage.

THE STEPS OF THE HACCP SYSTEM ARE:

1 First, **identify** potentially hazardous foods.

2 Then, **observe** those foods throughout your preparation, holding, and serving process, to **identify critical control points**.

3 **Establish control procedures** and monitor those critical points to guarantee safe handling of the food.

4 **Establish monitoring procedures** to adjust the process and maintain control.

5 **Establish corrective actions** to be taken when monitoring indicates that there is a deviation from an established critical limit.

6 **Establish effective record-keeping procedures** that document the HACCP system.

7 **Establish procedures to verify** that the HACCP plan or system is working.

The Steps of the HACCP System:
Explanation and Application of HACCP Principles

STEP 1: IDENTIFY POTENTIALLY HAZARDOUS FOODS

Hazard: Any biological, chemical, or physical property that may cause an unacceptable consumer health risk

Risk: A likelihood of a hazard

- Review description and charts of potentially hazardous foods (PHF).
- Review menu for potentially hazardous foods.

STEP 2: IDENTIFY CRITICAL CONTROL POINTS

A *critical control point* is defined as a point, step, procedure in which a food safety hazard can be prevented, eliminated, or reduced. Examples of critical control points (CCPs) may include but are not limited to: cooking, chilling, specific sanitation procedures, prevention of cross-contamination, and certain aspects of employee and environmental hygiene.

- Review critical items list.
- Observe foods throughout preparation, holding, and serving process.
- Review recipe procedures.
- Observe employee food handling and hand-washing practices.
- Observe use of sanitizer solutions and document proper concentrations.
- Conduct a pocket thermometer calibration demonstration.
- Chart the time/temperature of the cool down and/or reheating of a PHF.

STEP 3: ESTABLISH CONTROL PROCEDURES

Critical limits is defined as the criteria that must be met for each preventive measure associated with a CCP. Critical limits may be set for preventive measures such as temperature, time, physical dimensions, humidity, moisture level, water activity, pH, acidity, salt concentration, available chlorine, preservatives, or sensory information such as texture, aroma, and visual appearance.

- Incorporate control procedures into the written recipes; for example:

Chicken Breast
Minimum internal cooking temperature of chicken: 165°F

Oven temperature: _____°F

Time: rate of heating, cooling, reheating

- *Enforce* employee handwashing and hygiene practices.
- *Establish* illness policy for employees with flulike symptoms of diarrhea and vomiting.
- *Enforce* proper cleaning and use of sanitizer solutions.

STEP 4: ESTABLISH MONITORING PROCEDURES

Monitoring is a planned sequence of observations or measurements to assess whether a CCP is under control and to produce an accurate record for future use in verification. Examples of measurements for monitoring include

Visual observations
Temperature
Time
pH
Moisture level

Assignment of the responsibility for monitoring is an important consideration for each CCP. The person responsible for monitoring must also report a process or product that does not meet critical limits so that immediate corrective action can be taken. For example,

- Assign one person to make and test sanitizer solution each day.
- Assign responsibility for equipment temperature logs.
- Assign responsibility for food temperature logs for cooking, cooling, and reheating.

All records and documents with CCP monitoring are to be signed or initialed by the person doing the monitoring.

STEP 5: ESTABLISH CORRECTIVE ACTION

The HACCP system for food safety management is designed to identify potential health hazards and to establish strategies to prevent their occurrence. However, ideal circumstances do not always prevail. Therefore, when deviation occurs, corrective action plans must be in place to

- Determine whether food should be disposed of
- Correct or eliminate the cause of problem
- Maintain records of corrective action taken

Actions must demonstrate that the CCP has been brought under control. Individuals who have a thorough understanding of HACCP process, product, and plan are to be assigned responsibility for taking corrective action. Corrective action procedures must be documented in the HACCP plan.

STEP 6: ESTABLISH EFFECTIVE RECORD-KEEPING PROCEDURES

The associated records should be on file at the food establishment. Generally, such records include the following:

- Listing of the HACCP team members and assigned responsibilities
- Description of the food and its intended use/product description/specifications
- Listing of all regulations that must be met
- Ensure adequate environment, facilities, and equipment

- Monitor equipment with temperature logs
- Copies of flow charts from receiving to consumption
- Hazard assessment at each step in flow diagram (include calibration of equipment)
- The critical limits established for each hazard variable at each step:

Management	Equipment
Customers	Employees
Environment	Materials and supplies
Facility	Food production methods: handling from source to consumption

- Monitoring requirements for temperature, sanitation, finished product specifications, and distribution
- Corrective action plans when there is a deviation in policy, procedure, or standard CCP
- Procedures for verification of HACCP system

STEP 7: ESTABLISH PROCEDURES FOR VERIFICATION

Verification procedures may include:

- Establishment of appropriate verification inspection schedules
- Review of the HACCP plan
- Review of CCP records
- Review of deviations and dispositions
- Visual inspection of operations to observe whether CCPs are under control
- Random sample collection and analysis
- Review of critical limits to verify that they are adequate to control hazards
- Review of written record of verification inspections covering compliance, deviations, or corrective actions taken
- Review of modifications of the HACCP plan

THE HACCP
Food Safety Manual

STEP 1

Identify Potential Food Safety Hazards

A *hazard* is a food property that may cause an unacceptable health risk to your customers. Hazards may be biological, chemical, or physical.

Biological hazards include the presence of harmful bacteria, viruses, or other microorganisms. Biological hazards represent 93 percent of the incidences of food-borne illness.

- Bacterial Contaminants
- Survival of Bacterial Contaminants
- Cross-Contamination
- Other Biological Contaminants

Chemical hazards include toxins, heavy metals, and improperly used pesticides, cleaning compounds, and food additives. Chemical hazards account for 4 percent of foodborne illness occurrences.

Physical hazards include foreign objects that may cause illness or injury, for example, metal, glass, plastic, and wood.

To raise new questions, new possibilities, to regard old problems from a new angle requires creative imagination and marks real advances in science.

–Albert Einstein

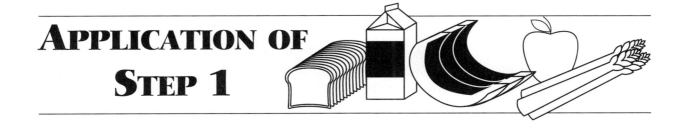

Identify Potential Food Safety Hazards

Examine the relationship between potentially hazardous foods (PHF) and foodborne illness.

Review the description and charts of potentially hazardous foods.

Review a menu for potentially hazardous foods.

Why Is Cleanliness Not Enough to Prevent Food Safety Hazards?

When customers visit a foodservice establishment they expect safe food. The only way they can judge food safety is by observation of the facility and its employees. They notice the employees' hands and uniforms. They also evaluate safety by visible cleanliness and general housekeeping appearances.

Cleanliness is critical if you want to be profitable, which means that customers return to your establishment. However, visual cleanliness does not assure safe food. Since the 1940s our food inspection has focused on the cleanliness of facilities, and yet foodborne illness has increased. In 1990, the U.S. Department of Agriculture reported an estimated 40 to 81 million cases of foodborne illness, resulting in more than 10,000 deaths each year.

Foodborne illness symptoms are very similar to the flulike symptoms of diarrhea, nausea, vomiting, and cramps. In many cases, people who assumed they had the flu were more likely experiencing a mild case of foodborne illness; thus the wide range of estimated cases of foodborne illness.

Foodborne illness is caused by bacteria, yeast, mold, viruses, parasites, and chemical substances. These contaminants originate from soil or through processing, transporting, or food handling. Improper handling of food accounts for a large percentage of the problem sources. Contamination during food handling is caused by people who prepare and serve the food: through the bacteria and viruses in the nose and mouth and on skin, and through the urine and feces. When foodservice personnel do not follow good personal hygiene and food handling practices, the pathogens are transmitted via foods to our customers, employees, and ourselves.

As the customer becomes better educated, so must we. Customer satisfaction and consumer confidence in the food supply have suffered because of a lack of basic knowledge of critical factors that cause foodborne illness. The food safety process begins by identifying the foods that are potentially hazardous. Then the critical points in processing or handling can be controlled, so that a hazard can be prevented, eliminated, or reduced. Finally, management must establish and enforce policies, procedures, and standards to ensure safety is achieved in all steps—from storage to preparation to presentation of food to the customer.

The Relationship Between Potentially Hazardous Foods and Foodborne Illness*

Many potentially hazardous foods have properties that support rapid bacterial growth and can cause consumer health risk. These foods, which include meats, cheeses, sauces, dairy products, eggs, beans, pasta, rice, and potatoes, can be found in most commercial food-service establishments.

Any food may be potentially dangerous to a customer when considering the biological, chemical, and physical hazards affecting a product.

Biological hazards are caused by pathogenic or disease-causing microorganisms commonly associated with humans and raw products.

Chemical hazards occur when chemical substances, either naturally occurring or added during processing, enter into the food supply process.

Physical hazards occur when hard foreign objects that are not intended to be part of the food get into the food through contamination and/or poor procedures.

The major causes of foodborne illness are bacteria and toxins in consumed food, which are included in the following list:

Biological Hazards
- Bacteria
- Viruses
- Parasites
- Fungi
- Molds
- Yeasts

Physical Hazards
- Glass
- Toothpicks
- Nonedible garnishes
- Metal shavings
- Nails, staples, etc.

Naturally Occurring Chemicals
- Fish Toxins
  - Ciguatera
  - Scombroid poisoning
- Plant Toxins

Added Chemicals
- Pesticides
- Additives and preservatives
- Toxic metals
- Foodservice chemicals

*Joan Hohn, Hospitality Project Manager, Consultant, Lakewood, Colorado.

How do these agents get into food? Bacteria and toxins are transferred to the food, then on to the customer, by people, poor food handling practices, and/or equipment. Some foods are naturally contaminated, such as raw meats, unwashed produce, and certain fish species and plants. When a susceptible customer ingests contaminated food, the action results in the ill health of the person.

What is the most common source of illness? Biological hazards are the source of the majority of outbreaks. If the right conditions of time, temperature, moisture, and pH exist for growth, they will thrive in many of our favorite foods (potentially hazardous foods). Pathogenic organisms like to grow in the same environments as we do; at room temperature (if given enough time), where they have moisture and a neutral pH. Potentially hazardous foods provide the right medium (food) to grow, given the proper environment (time and temperature). Once foods contaminated with pathogenic bacteria and/or viruses are eaten, the pathogens grow within the body and poison the consumer with their toxins.

What can be done to prevent foodborne illness? Intensive training can alert food handlers to the sources of danger. Understanding, consistent corrective action, and staff education will minimize the likelihood of foodborne illness injuring your customers.

HACCP gives us guidelines for preventing foodborne illness. In general, we must alert managers and food handlers continually to question and control the food preparation process from beginning to end:

- Food source, soil, or water conditions in which the food is grown
- Processing, shipping, receiving, and storage
- Food handler health and hygiene
- Kitchen sanitation and equipment efficiency
- Cooking, holding, cooling, reheating, and serving practices

Sources of Foodborne Organisms in Foodservice Establishments

HUMANS
(Workers)

Nose and Throat
Hands
Infections
Feces
Clothing

FOODS
(of Animal Origin)

Poultry
Meat
Eggs
Fish/Shellfish

(of Plant Origin)

Soils
Water

For Growth of Foodborne Disease, Organisms Need:

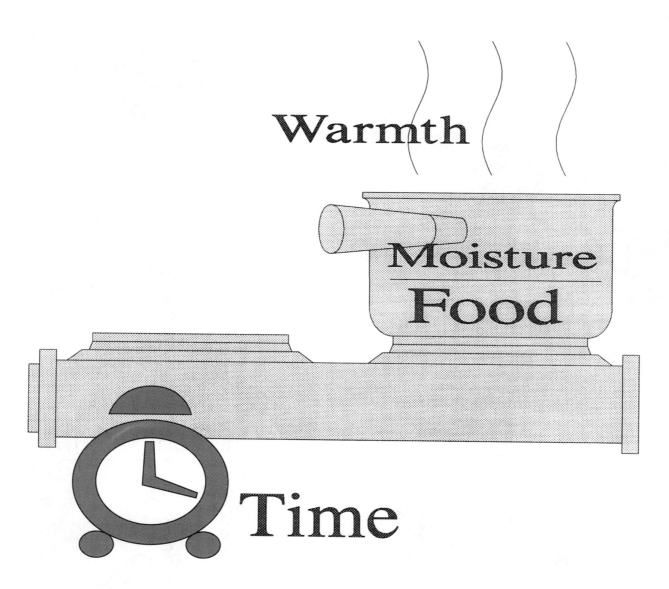

Warmth

Moisture

Food

Time

A Focus on Foodborne Disease Factors

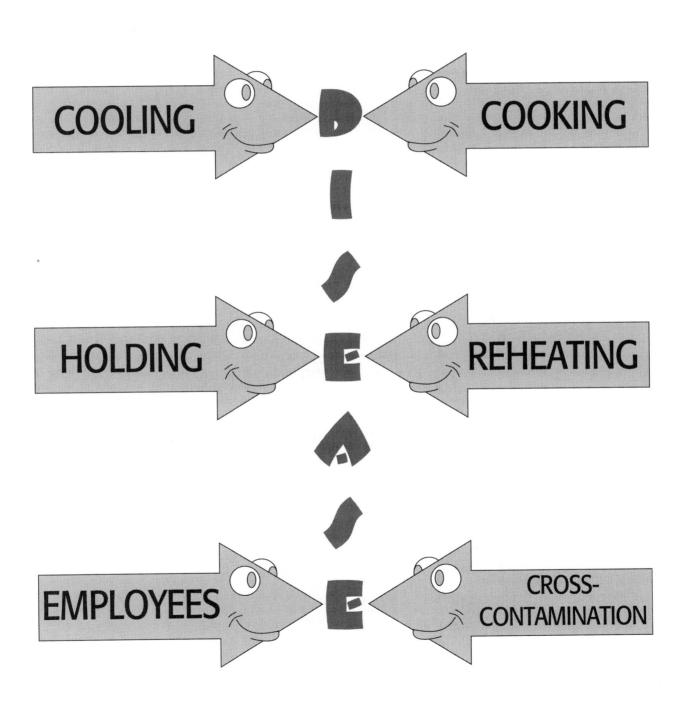

Potentially Hazardous Foods

"Potentially hazardous foods" means any natural or synthetic food or food ingredient that supports the rapid growth of infectious or toxigenic microorganisms or the slower growth of *C. botulinum.*

A food is potentially hazardous if it is:

- Of an animal source, such as meat, milk, fish, shellfish, edible crustacea, poultry, or contains any of these products
- Of plant origin and has been heat treated
- Any of the raw seed sprouts

The following is a partial list of specific food products that have been classified as potentially hazardous:

1. *Bacon*—If it has not been fully cooked.

2. *Beans*—All types of cooked beans.

3. *Whipped butter*—Because of the apparent reduced microbiological safety factor created in whipping.

4. *Cheese*—Soft, unripened cheese. Ripened, low-moisture hard cheese, such as wheels, flats, blocks, longhorns, or cheddar produced from pasteurized milk, when waxed or packaged in shrink-wrap with the wax or packaging intact, can be safely shipped or stored for a short period of time without refrigeration, but this is not recommended. If wheels or flats are cut and repackaged for display and/or sales, thereby exposing interior surfaces to possible contamination, the cut portions as well as the remaining cheese must be held under refrigeration.

5. *Coffee creaming agents*—All nondairy coffee creaming agents in liquid form, except aseptically processed ultrahigh temperature (UHT) liquid coffee creaming agents.

6. *Eggs*—Fresh eggs in shells, fresh eggs with outer shell removed, peeled hard-boiled eggs, and hard-boiled eggs with intact shells which have been hard-boiled and then cooled in liquid.

7. *Garlic*—Unrefrigerated, fresh garlic in oil products provide the anaerobic (oxygen-free) environment required for *C. botulinum.*

8. *Onions*—Cooked and dehydrated onions that have been reconstituted.

9. *Pasta*—All types that have been cooked.

10. *Pastries*—Meat, cheese, and cream filled.

11. *Pies*—Meat, fish, poultry, natural cream, synthetic cream, custard, pumpkin, and pies that are covered with toppings that will support microbial growth.

12. *Potatoes*—Baked, boiled, or fried.

13. *Rice*—Boiled, steamed, fried, Spanish, and cooked rice used in sushi.

14. *Sauces*—Hollandaise and other sauces containing potentially hazardous ingredients. If these are held in the temperature range of 45° to 130°, they must be discarded within four hours of preparation.

15. *Sour Cream*—If the pH is above 4.6 and/or the sour cream is combined with other food products.

16. Soy protein—Tofu and other moist soy protein products.

17. *Seed sprouts*—All types.

Foods that are *not* potentially hazardous are:

1. Hard-boiled eggs with shells intact, which have been air dried.

2. Foods with a water activity (A_w) value of 0.85 or less.

3. Foods with a measurement of acidity (pH) of 4.6 or below.

4. Foods that have been adequately commercially processed and remain in their unopened hermetically sealed containers.

5. Foods for which laboratory evidence (acceptable to the regulatory authority) demonstrates that rapid and progressive growth of infectious and toxigenic microorganisms or the slower growth of *C. botulinum* cannot occur.

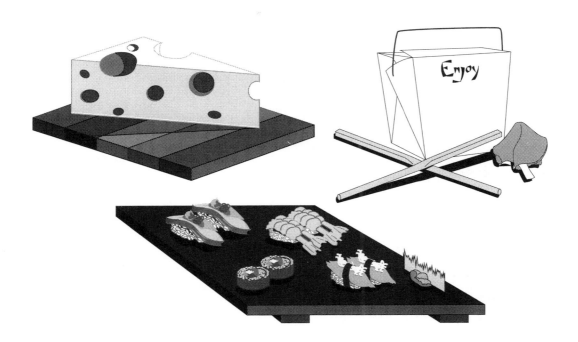

Cooked or Raw Animal Products:
Meats, Dairy, Eggs

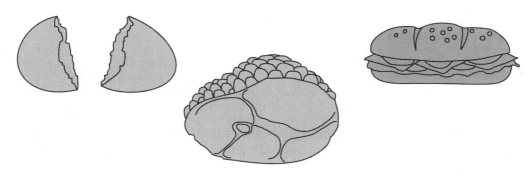

Cooked Vegetables & Starches & Raw Seed Sprouts

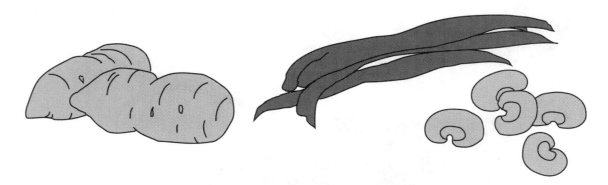

POTENTIALLY HAZARDOUS FOODS

 BEANS / RICE / PASTA

PIES / PASTRIES / EGGS

 POTATOES / SEED SPROUTS

SOY PROTEIN / MEATS

CHEESE / WHIPPED BUTTER

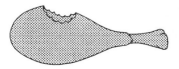 CHICKEN / SHELLFISH

DAIRY / NONDAIRY AGENTS

FRESH GARLIC IN OIL

New Generation Refrigerated Foods*

New generation refrigerated foods eliminate some of the preparation steps of foods usually prepared in a foodservice operation. These food items are packaged to extend shelf life. While the packaging inhibits growth of spoilage organisms, it may promote growth of pathogenic bacteria such as *C. botulinum* or *Listeria* if the foods are time/temperature abused or served beyond the recommended "Use-by" dates. Receiving and storage temperatures are critical for these products.

New generation foods are sous vide and MAP foods, as explained in the following paragraphs.

TYPES OF PACKAGING

Modified atmosphere packaged (MAP) foods: Food is partially processed or lightly cooked before being put into a pouch or other container and sealed. The atmosphere in the package is usually a mix of oxygen and carbon dioxide. These foods should be received and stored at temperatures of 40°F (4.4°C) or lower.

Sous vide: Food is put into a package raw and sealed under vacuum; then it is heat treated. These foods should be received and stored at temperatures of 40°F (4.4°C) or lower.

Aseptic packaging: This is a method whereby both the food product and the packaging are sterilized before filling and sealing. These products do not have to be refrigerated until after being opened.

Ultrapasteurized: These dairy products are not aseptically packaged, and must be received and stored at temperatures of 45°F (7.2°C) or lower.

UHT: These dairy products are aseptically packaged and ultrapasteurized. They do not need to be refrigerated before opening.

EXAMPLES

New generation foods that require proper refrigeration at receiving and storage are as follows: Refrigerated entrees, prepared salads, fresh pasta, soups, sauces, gravies, and cooked or partially cured meats and poultry dishes.

CONCERNS AND BENEFITS

Because these new generation refrigerated foods are minimally or partially processed, they present both benefits and concerns.

Benefits
- High quality
- Labor saving
- Menu flexibility
- Convenience
- Extended shelf life

---

*Adapted with permission from *Managing a Food Safety System* seminar. Copyright © 1992 by the Education Foundation of the National Restaurant Association.

Concerns

- Refrigeration may be the only barrier to pathogenic growth.
- Psychrophilic bacteria such as *Listeria* and some strains of botulism may not be killed.
- Competing spoilage organisms may be killed. Spoilage organisms inhibit the growth of pathogenic bacteria by competing with them. You can see evidence of spoilage organisms, whereas the presence of disease-causing organisms is not noticeable.
- Pathogens and their spores may not be destroyed.
- An anaerobic condition is created, favoring the growth of pathogens such as *C. botulinum* or *Listeria*.
- Cooking may make food more favorable to pathogen growth.

Therefore, these recommendations should be followed:

- Do not sell MAP (modified atmosphere package) foods at retail level.
- Use temperature indicators to ensure that foods were not temperature abused in transit.
- Practice rigorous time/temperature control.
- Train employees.
- Read all labels carefully to make sure you follow the temperature requirements for each product.

Fish and Shellfish, Unique Food Safety Concerns*

Fish and shellfish are nutritious foods that constitute desirable components of a healthy diet. Most seafood available to the U.S. public is wholesome and unlikely to cause illness in the consumer. Nevertheless, there are areas of risk. Traditional food safety regulation techniques do not adequately address the unique concerns involved in seafood safety. CDC (Centers for Disease Control) data indicate that most seafood-related illnesses result from certain natural toxins in finfish and from viruses in molluscan shellfish consumed raw or partially cooked. Other risks include parasites, such as *Anisakis,* consumed with raw or undercooked fish. Finfish are also generally regarded as being much more perishable than other foods and are implicated in the formation of scombrotoxin during decomposition.

MICROORGANISMS AND PARASITES

Seafood, like any food item, has the potential to cause disease from viral, bacterial, and parasitic microorganisms under certain conditions. These agents are acquired from four main sources: fecal pollution, the natural environment, processing, and preparation. With the exception of foods consumed raw or partially cooked, the reported incidence of seafood-related microbial diseases is low.

Consumers should be informed of the risks of eating raw seafood, particularly shellfish (especially bivalve mollusks, oysters, clams, and mussels), and be advised to cook seafood sufficiently to destroy parasites and bacterial contaminants.

PARASITE DESTRUCTION

Freezing is required before service to destroy parasites in raw, marinated, or partially cooked fish, other than molluscan shellfish.

- In the freezer for seven days at −4°F (−20°C) or below
- In a blast chiller for 15 hours at −31°F (−35°C) or below

NATURAL TOXINS

Naturally toxic fish and shellfish are not distinguishable from nontoxic animals by sensory inspections, and the toxins are not destroyed by cooking or processing. Except for scombroid fish poisoning, natural intoxications are both highly regional and species associated. Such toxins are present in the fish or shellfish at the time of capture. Scombroid poisoning is due to histamine produced by bacteria multiplying on the fish that are time/temperature abused after capture.

Scombroid poison should be controlled primarily by rapid chilling on the fishing vessel and by maintenance of refrigeration temperatures throughout distribution. Some implicated fish include tuna, mahi mahi, bluefish, sardine, mackerel, amberjack, and abalone.

Ciguatera is a sometimes severe disease caused by consuming certain species of fish from tropical waters usually associated with islands and reefs. The disease is most common in the Caribbean and Pacific islands. Some outbreaks caused by imported fish occur in southern Florida and in other states. Implicated fish include grouper, barracuda, snapper, amberjack, mackerel, and triggerfish.

Seafood Training Program, National Academy of Sciences Report on Seafood Safety, National Academy Press, Washington, D.C.: 1991.

At present there are no effective control systems in place for the prevention of ciguatera. Ciguatera is best controlled by not purchasing or eating particular species of fish from areas where ciguatera warnings have been issued.

TEMPERATURE CONTROL AND SHELF LIFE*

All of the changes that cause seafood spoilage are affected by temperature. High temperatures speed spoilage, and low temperatures slow spoilage. For many seafood species, increasing temperature from 32°F to 40°F doubles the rate of spoilage and cuts the shelf life in half.

Sanitation is also important. Contamination of seafood by bacteria from dirty ice, containers, and surfaces can increase the number of bacteria and speed spoilage. Keeping seafood handling and storage equipment clean reduces bacterial contamination and slows spoilage.

The approximate shelf life for fresh fish is as follows.

Holding Temperature	High-Quality Shelf Life	Edible Shelf Life
42°F	3½ days	6 days
32°F	8½ days	14 days
30°F	10½ days	17½ days
29°F	12 days	20 days

Seafood shelf life relates directly to storage time and temperature. Your supplier cannot guarantee a shelf life for a seafood product without knowing the catch date and the temperature history. Ideally, time-temperature monitors should accompany seafood from the fishing vessel to the retail store, but this is rarely feasible.

TEMPERATURE RECOMMENDATIONS

1. Train staff to inspect seafood upon delivery. They should have the ability to accept or reject any shipment. Rejection may be for odor, appearance, or temperature.

2. Reject seafood products with a temperature above 35°F and return them to the shipper.

3. Time-temperature monitors should accompany all seafood shipments. Monitors can be color-changing, temperature-sensitive badges or recording thermometers.

4. Require a statement indicating when the processor attached the time-temperature monitor to the product.

**Retail Seafood Temperature Control,* University of California Extension, Sea Grant Extension Program Publication, Robert J. Price, Ph.D., Extension Seafood Technology Specialist, University of California, Davis, 1990.

5. Reject products when temperature records are not available from the shipper.

6. Thaw frozen seafood under refrigeration. Thaw only enough seafood to sell in a 24-hour period.

7. Store fresh seafood at 29°–32°F. A double self-draining pan system using ice will help maintain the temperature.

8. Do not store or display red meat and seafood in the same case or pan. These products have different storage temperature requirements.

9. Display unpackaged fresh seafood at 29°–32°F. Use a mechanically refrigerated display case equipped with an accurate thermometer. Refrigeration coils should be at the top of the case, not at the bottom.

10. Routinely check display case and seafood temperatures. Check seafood with a sanitized, calibrated thermometer.

Our Changing Food Chain and Food Safety Implications

Why does food safety seem to be of great concern today? It is because our evolving food chain has created new food safety problems that didn't exist 10 years ago. Some of these problems are a result of our demand for low food prices.

POULTRY AND EGGS

The factory raising of poultry creates cramped environments, allowing salmonella to spread easily through the flocks. Then, in large processing plants the birds are mechanically handled, and feces from intestines get on the flesh and contaminate the birds. It is estimated that nearly 60 percent of these chickens are contaminated with salmonella. We have to assume that each piece is contaminated and handle the poultry in such a manner as to prevent cross-contamination. Poultry must be cooked to an internal temperature of 165°F to ensure that the salmonella organisms are killed.

Only recently have we had to be concerned about salmonella in fresh whole eggs. We have had to change the way we handle eggs, keeping them chilled until ready to use. To reduce risk, the use of pasteurized eggs for cooking has increased dramatically. Many classic egg recipes, such as hollandaise sauce and Caesar salad, have had to be modified to eliminate the use of fresh eggs.

GROUND BEEF

For years we wanted low milk and beef prices. As a result, we now have *E. coli* 0157:H7. To increase milk production, farmers gave their dairy cows antibiotics. After years of antibiotic use (no longer used for milk production), a new strain of *E. coli* evolved. However, this wasn't known until the dairy cows were put into the food chain. When milk subsidies were lowered or dairy cows got old, they were sent to slaughter where they were often used in ground beef.

In processing plants, cows are mechanically slaughtered. Sometimes the intestines rupture and *E. coli* from the intestines gets on the carcass. The carcass is hosed off, but this does not remove all the microscopic bacteria. When the meat is ground, the bacteria become mixed in. Unless all ground meat is cooked to 155°F consumers are at at risk to *E. coli* 0157:H7 illness.

PRODUCE

Today we enjoy a year-round supply of a great variety of produce. To achieve the required supply system, produce is imported from Third World countries. These countries which do not have the sanitary standards of the United States, use human fertilizer for growing produce. Even in the United States, field workers are not provided with adequate rest room and hand-washing facilities.

Therefore, it is essential that all produce be washed before processing, cutting, and cooking to remove the bacteria from the surface. This will reduce the likelihood of transferring organisms such as *Hepatitis A, Clostridium perfringens, Shigella, E. coli,* and *Bacillus cereus* through soil and human carriers.

EMERGING PATHOGENS

Two emerging pathogens are causing problems because they can grow slowly at refrigeration temperatures. *Listeria* and *Yersinia* grow at 45–40°F. Although the disease is rare, *Listeria* is of particular concern, because it causes more deaths per incidence of illness than any of the other foodborne illnesses. To control its effect, avoid raw milk and cheese made from unpasteurized milk; follow KEEP REFRIGERATED, SELL BY, and USE BY labels; thoroughly reheat frozen and refrigerated products; sanitize equipment before preparing foods to be eaten raw; and follow proper hand-washing procedures.

Foodservice operators can significantly reduce the risk of foodborne illness by concentrating their efforts in three areas:

- Enforce employee health and hygiene practices.
- Follow proper temperature control and monitoring during cooking, cooling, holding, and reheating foods.
- Follow recommended use and concentrations of sanitizers (discussed in Step 2).

These three significant procedures to reduce foodborne illness are discussed in greater detail in the following HACCP steps. For more information about foodborne illness, refer to the "Foodborne Illness" section of this book.

HAZARD — Unacceptable contamination, unacceptable microbial growth, unacceptable survival of microorganisms of concern to food safety, or persistence of toxins.

RISK — Probability that a condition or conditions will lead to a hazard.

SEVERITY — Seriousness of the consequences of the results of a hazard.

IDENTIFY HAZARDS*
(contamination, survival, growth)

- ## Review Menu and Recipes

- ## Observe Employees

- ## Seek Additional Facts (ask questions)

- ## Measure Temperatures

- ## Test Foods

- ## Review Records

*Adapted with permission from Managing a Food Safety System seminar. Copyright © 1992 by the Education Foundation of the National Restaurant Association.

Identify Hazards

Numerous issues have to be addressed during hazard analysis. These relate to factors such as ingredients, processing, distribution, and intended use of the product. These issues include whether a food contains sensitive ingredients that can create a microbiological, chemical, or physical hazard, and whether sanitation practices in use can introduce these hazards to food.

This point in hazard analysis consists of asking a series of questions that are appropriate at each step in the food flow. You will want to identify situations in the foodservice operation where there is the possibility of unacceptable contamination, unacceptable microbial growth, survival of microorganisms of concern to food safety, or the persistence of toxins.

Review Menu and Recipes

- Identify the potentially hazardous foods.
- What foods have been associated with outbreaks in the past?
- Which foods are produced in large quantities?
- Are foods prepared in advance? How long?
- Which foods require multi-preparation steps?
- Are foods going through cooking, cooling, and reheating cycles?
- Is there any new cooking/processing technique used, such as sous vide, cook-chill, or others?
- Are raw eggs used in recipes?
- Is raw seafood served? Are other raw foods of animal origin served?
- Are wild mushrooms served? What is the source?
- What recipes incorporate leftovers? How are they being handled?
- Are sulfites used? How?

Observe Employees

- Are employee hands washed? When?
- What are the policies on illness, cuts, and burns?
- Is there an employee food safety training program? Describe.
- What food items on the menu are likely to be extensively handled by employees?
- Where are salad ingredients washed and prepared?
- What procedures are being followed for cleaning and sanitizing of equipment and utensils?
- Are sanitizing solutions and wiping cloths properly used?

Seek Additional Facts

- Is catering being done outside of the establishment?
- Are you aware of recommendations about immune-compromised individuals?
- Ask follow-up questions to obtain further details about preparation.

Measure Temperatures
- How do you tell when foods have finished cooking?
- How do you determine when the following foods have finished cooking: pork, poultry, beef, rare roast beef, items containing ground beef?
- Do you know how long it takes to cook foods placed in the walk-in refrigerator or other refrigerators?
- What are the hot-holding temperatures?
- What are the refrigeration temperatures?
- How do the employees know how long a food has been stored?
- How are foods thawed?
- Do employees check the accuracy of food thermometer? How?

Test Foods for Acceptance or Risk

Larger chains often test foods to determine contamination of certain foods, such as salmonella in veal, or the acidity of a salad dressing or spaghetti sauce. This can be useful in knowing how a food is to be handled or risk associated with certain foods. Laboratory tests can be useful for:

- pH of foods
- Microbial contamination
- Product tampering
- Product quality assurance

Review Records
- What are the primary food supplier sources?
- What are the product specifications?
- Are shellfish tags kept? For how long?
- What forms for documentation are used?
- Are time and temperature logs maintained?
- What corrective actions have been taken? Has documentation been kept?
- Are records of employee training kept?

Estimate Risks

Risk is the probability that a condition or conditions will lead to a hazard. Each type of food operation and each establishment presents different risks to the consumer. Factors in operational risk include the type of customers served, the menu, the size of the operation, the type of service, and employee training. An operation that establishes an HACCP system will reduce its risk. However, a foodborne illness outbreak could occur almost anywhere, because not all hazards are completely controllable by the operator and some customers are more susceptible than others.

Customers

- Young, elderly, and immuno-suppressed customers are more likely to become ill or more seriously ill if they eat contaminated or undercooked food.
- Establishments that serve a large number of customers are more likely to have an outbreak occur.
- Operations that offer take-out and delivery have a higher risk, because they lose control over how the food is handled once it leaves the foodservice establishment.

Suppliers

- When products are purchased through state approved sources, risk is reduced, because they have to meet certain standards when they are inspected.
- Risk can be reduced when you require your seafood supplier to submit a Certificate of Conformance. This certificate is a contract with your supplier stating that there will be no species substitution and that the seafood is from safe and approved waters.

Size and Type of Operation

- Establishments selling a large number of menu items have a higher risk than a facility selling fewer menu items.
- A quick-service operation has a lower risk than a full-service or institutional type of operation.
- An operation that serves more customers, a greater number of menu items, or types of menu items for which it does not have the proper equipment or facilities, has a higher risk.
- Operations with complicated or multistep recipes have greater risks than operations with simple-process recipes and cook-and-serve products.

Employees

- Foodservices that have employees trained in proper handling and preparation of food have lower risks.
- Operations that maintain good sanitation and housekeeping practices have lower risks.

ESTIMATE RISKS*

1. Customers
- Young, Elderly, Immuno-suppressed
- Number of Customers
- Delivery and Take-out

2. Suppliers
- State Approved Sources
- Certified Seafood Shippers

3. Size and Type of Operation
- Menu and Pace of Service
  (Quick-service, Full-service, Institutional)
- Proper Equipment & Facilities
- Complicated or Multi-Step Recipes

4. Employees
- Trained in proper handling and preparation of food

*Adapted with permission from Managing a Food Safety System seminar. Copyright © 1992 by the Education Foundation of the National Restaurant Association.

Identify Potentially Hazardous Foods

Review the menus on pages 27–29. Identify the potentially hazardous foods (PHF).

Next assess menu risks for:

1. Naturally contaminated foods

2. Foods implicated in previous outbreaks

3. Quantity preparation

4. Multistep preparation

5. Cooking, cooling, and reheat cycles

6. Use of leftovers

Are there some recommended menu or procedural changes that could be suggested to reduce risk?

> *S*ooner or later the idea here put forth will conquer the world, for with inexorable logic it carries with it both the heart and the head.
>
> *–Albert Schweitzer*

Breakfast Specials

Mon: Chicken, Broccoli, Omelette W/ Toast
Tues: Chipped Beef Gravy & Biscuit, 2 Links or Bacon
Wed: Huevos Rancheros
Thur: Sicilian Frittata
Fri: Cheddar, Bacon Skillet
Daily: Belgian Waffle

Pancake: Bakery:

Mon: Buttermilk   Blueberry Coffee Cake
Tues: Blueberry   Pecan Rolls
Wed: Buttermilk   Cherry Coffee Cake
Thur: Blueberry   Cinnamon Rolls
Fri: Applesauce   Bread Pudding

Regular Breakfast Fare

Eggs ....
Hash Browns ....
Bacon ....
Sausage ....
Hot Cereal ....
Cold Cereal ....
Toast ....
Large Sausage Patty
Texas French Toast

Lunch Grill

Hamburgers, French Fries, BLTs, Clubs, Fish or Chicken Filet, Tuna Melt or Patty Melt are available to order.

Now serving;
    Quesadilla ....
    Smothered Beef Burrito...
Smothered Chicken Burrito ....

Food For Thought

You will notice that nearly every food item on this cafeteria menu is a potentially hazardous food. These items contain proteins of eggs, meats, dairy, and cooked starches and vegetables.

Most of the bakery or dessert items are not potentially hazardous, except for the cream pies, frozen yogurt, and bread pudding.

Salad Bar:
Salads to go: Chef, Cobb, Tuna, Crab, Spinach, Fruit & Cheese, Chicken Breast

Fast Food

Tues. & Thur. — ***Potato Bar***

Soup Lover's Delight

Mon: Tortilla
     Swiss Potato

Tues: Beef Barley
     Cream of Broccoli

Wed: Chicken Noodle
     Navy Bean

Thur: Minestrone
     Beer Cheese

Fri: Chicken & Corn

Entrées:

Monday:
- Swiss Steak
- Vegetarian Lasagna

Tuesday:
- Chicken Fried Steak
- Baked Cod ....

Wednesday:
- Super Nachos ....
- Southern Fried Chicken

Thursday:
- Turkey / Dressing ...
- Salisbury Steak......

Friday:
- Lemon Chicken & Rice
- Ginger Beef & Rice

Two Vegetables, Rolls and Butter available with the above entrées

Deli
(Made to Order)
Ham, Turkey, Roast Beef, Salami, Braunsweiger, Tuna, & Cheeses

Mon: Italian Hoagie
Tues: Hoagie Club
Wed: Corned Beef & Swiss
Thur: Greek Pita Pocket
Fri: Roast Beef, Turkey, Ham & Cheese

Dessert Specials

- Layer Cake ....
- Fruit & Cream Pies ....
- Fudge Brownies............
- Cookie ....
- Fresh Fruit Compote .....
- Frozen Yogurt..............

Appetizers

Wild Mushroom Warmed with a Dried Tomato and Basil
Vinaigrette, Chevre Crouton, Served on Spring Greens

Caesar Salad of Romaine and Radicchio Leaves Prepared
in the Traditional Manner at Your Table

Select Oysters of the Season Served on the Half Shell
with Champagne Shallot Mignonette

Carpaccio of Buffalo and Veal Presented with Lime,
Shallots, Asiago Cheese, Capers, and Cracked Pepper

Entrées

Salad Nicoise of Yellowfin Tuna Charred Rare Served with Grilled Potatoes,
Calamata Olives, Crispy Fried Capers, and Sweet Soy

Roasted Breast of Chicken Stuffed with Gorgonzola Cheese
and Prosciutto Ham Served on a Warm Agnalotti Salad

Grilled Red Rock Cod Served on Sourdough Bread
with Caper Aioli and Fried Sweet Potatoes

Red Deer Carmelized with Cinnamon Jam, Sweet Potato Gratin,
Spaghetti Squash Cake, and Casis Sauce

Filet of Beef Charred with Chilies and Soy with a Parsnip
Potato Cake and Confit of Leeks

Zita Pasta Tossed with Curry Cream, Peanuts
with a Roasted Pepper Vinaigrette, and Red Onion Confit

BREAKFAST
(Please Circle Your Selection)

CHILLED JUICES & FRUIT

Citrus Sections   Orange Juice
Apple Sauce   Apple Juice
Banana   Grape Juice

CEREALS

Hot Cereal   Puffed Rice
Corn Flakes   All Bran
Rice Krispies   Special K
Raisin Bran   Shredded Wheat
Sugar Smacks

HOT SELECTIONS

Scrambled Eggs   *Bacon*
Low Cholesterol Eggs   *Sausage Links*
Pancakes
Cheese Omelette

Toast (White or Wheat)
Fresh Bakery Item

BEVERAGES

Coffee   *Buttermilk*
Decaf Coffee   *2% Milk*
Hot Tea   *Skim Milk*
Herb Tea   *Chocolate Milk*
Iced Tea   *Iced Herb Tea*
Extra Sugar   *Cream*

NAME_____ ROOM_____

LUNCH
(Please Circle Your Selection)

Minestrone Soup

MAIN COURSE

*Veal Picata: Slices of Veal
with Tangy Lemon Caper Sauce
and Pasta*

*Manicotti: Cheese Stuffed Pasta
Baked in Rich Tomato Marinara*

Stuffed Tomato Crown: Tomato
Stuffed with *Tuna Salad,*
Crackers and Cold Relishes

Hot Entree: Served with Selected
Vegetable(s) and Dinner Roll

TO COMPLIMENT YOUR MEAL

Tossed Spinach Salad with
Poppy Seed Dressing
Brownie

BEVERAGES

Coffee   *Buttermilk*
Decaf Coffee   *2% Milk*
Hot Tea   *Skim Milk*
Herb Tea   *Chocolate Milk*
Iced Tea   *Iced Herb Tea*
Extra Sugar   *Cream*

NAME_____ ROOM_____

DINNER
(Please Circle Your Selection)

Cream of Broccoli Soup

MAIN COURSE

*Baked Halibut Almandine: Served
with Rice Pilaf*

*Baked Meat Loaf: Served with
Mashed Potatoes & Gravy*

*Vegetable Pita Pocket: Pita Bread
Stuffed with a Blend of Green Beans,
Mushrooms, and Cheese (Baked)*

Hot Entree: Served with Selected
Vegetable(s) and Dinner Roll

TO COMPLEMENT YOUR MEAL

Fresh Fruit Salad
Ice Cream

BEVERAGES

Coffee   *Buttermilk*
Decaf Coffee   *2% Milk*
Hot Tea   *Skim Milk*
Herb Tea   *Chocolate Milk*
Iced Tea   *Iced Herb Tea*
Extra Sugar   *Cream*

NAME_____ ROOM_____

Potentially hazardous foods are italicized.

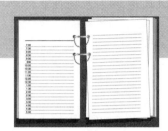

HACCP Project

In the following exercise, you will work through the first HACCP step.

1. Identify potentially hazardous foods (PHF).

2. Identify critical control points.

3. Establish control procedures to guarantee safe food.

4. Establish monitoring procedures.

5. Establish corrective action.

Start by reviewing the Fried Chicken Breast recipe, then go to Step 1 on this page.

FRIED CHICKEN BREAST

Items:

Boneless Chicken Breast
Flour
Salt
White Pepper
Eggs
Milk

Preparation:

1. *Mix* flour, salt, and pepper.
2. *Mix* egg with milk, blend well.
3. *Dip* chicken in egg wash, then in flour.
4. *Deep fat fry* in 375°F for approximately 4 minutes.
5. Remove, place on wire rack in sheet pan.
6. *Bake* in 375°F oven for approximately 20 minutes until done.
7. Remove and chill.
8. *Reheat* to serve.

HACCP Step 1: Identify Potentially Hazardous Foods (PHF)

FRIED CHICKEN BREAST

A. Review Potentially Hazardous Foods on pages 10–13.
    A food is potentially hazardous if it is (1) cooked or raw animal products such as meat, milk, fish, shellfish, edible crustacea, poultry, or contains any of these products; (2) cooked vegetables or starches; or (3) raw seed sprouts.

B. Does the Fried Chicken Breast meet the criteria for a potentially hazardous food (PHF)?

C. List the PHFs used in the Fried Chicken Breast recipe.

    1.

    2.

    3.

Answers are on page 32.

ANSWERS:

HACCP Step 1: Identify Potentially Hazardous Foods (PHF)

FRIED CHICKEN BREAST

A. Review Potentially Hazardous Foods on pages 10–13.
A food is potentially hazardous if it is (1) cooked or raw animal products such as meat, milk, fish, shellfish, edible crustacea, poultry, or contains any of these products; (2) cooked vegetables or starches; or (3) raw seed sprouts.

B. Does the Fried Chicken Breast meet the criteria for a potentially hazardous food (PHF)?　Yes

C. List the PHFs used in the Fried Chicken Breast recipe.

     1. Chicken

     2. Egg

     3. Milk

STEP 2

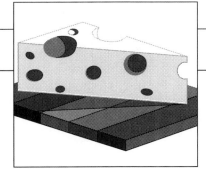

Identify Critical Control Points

A critical control point is defined as a point, step, or procedure at which a food safety hazard can be prevented, eliminated, or reduced. Examples of critical control points (CCPs) may include, but are not limited to:

Employee and environmental hygiene

Prevention of cross-contamination

Specific sanitation procedures

Cooking

Chilling

*C*risis:
the point at
which something
gets either
much better or
much worse.

–Author Unknown

APPLICATION OF STEP 2

Identify Critical Control Points

Review critical items list.

Observe hand-washing practices.

Observe food handling, potential cross-contamination, and use of sanitizer solutions.

Observe foods throughout preparation, holding, and serving process.

Review recipe procedure.

Conduct a probe thermometer calibration demonstration.

Chart the time/temperature of a cool down and/or reheat of a potentially hazardous food (PHF).

CRITICAL ITEMS

 Personal hygiene

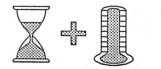 Time / temperature

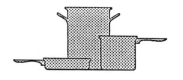 Cooking, cooling, reheating, holding

 Preparation ahead of time

 Cross-contamination

Critical Items

Critical items are the problems and practices that are the leading factors in outbreaks of foodborne illness. When these are properly controlled in the food preparation process, food safety hazards can be prevented, reduced, or eliminated, because bacteria will not have an opportunity to grow. Critical items in preventing foodborne illness are:

- Time/temperature
- Cooking, cooling, reheating, holding
- Preparation ahead of time
- Cross-contamination
- Personal hygiene

In controlling critical items,

- Foods must be kept at proper temperatures for appropriate periods of time.
- Proper heating and cooling methods must be used.
- There must be proper management of potentially hazardous foods prepared ahead of time, especially foods in bulk.
- You must prevent the movement of bacteria from one area to another (cross-contamination).
- You must practice appropriate personal hygiene.

A *critical control point (CCP)* is a step in the product-handling process whereby controls will reduce, or eliminate or prevent hazards. A critical control point is the "kill" step in which bacteria are killed by cooking or the "control" step that prevents or slows their growth, such as proper chilled storage or hot holding.

Examples of CCPs may include, but are not limited to:

- Cooking, reheating, and hot-holding stages
- Chilling, chilled storage, and chilled display stages
- Receiving, thawing, mixing ingredients, and other food-handling stages
- Specific sanitation procedures
- Prevention of cross-contamination
- Employee and environmental hygiene

Potential Food Safety Hazards — Critical Items

The following lists give examples of potential food safety hazards.

Improper Hot or Cold Storage
- Foods stored at improper temperatures
- Foods thawed at room temperature

- Coolers and display units without thermometers
- Poor cooling practices: overloading refrigerator units
- Foods cooled in large volumes and large containers
- Inadequate reheating of foods
- Hot display cases without thermometers
- Storage of food in improperly labeled containers
- Transport of food at improper holding temperatures

Cross-Contamination

- Storage of raw foods with ready-to-eat foods
- Employee practices leading to cross-contamination
- Preparing raw foods at same time and in same work area with cooked foods
- Using hands instead of utensils for raw foods that will not be further cooked
- Failure to clean equipment properly
- Failure to adequately protect food from contamination
- Employees working who have flulike symptoms
- Improper storage of refuse in food preparation areas

Other Hazards

- Improper or inadequate cleaning and sanitation practices
- Inadequate use of sanitizers
- Poor food preparation and handling practices
- Utensils or food contact surfaces made from improper or unimproved materials
- Inadequate documentation and record keeping
- Improper storage of chemicals or personal items

Review "General Operational Practices That Pose High Risk" (see page 39). Indicate which critical items are most likely implicated with each operational practice. Complete the assignment for the first nine "Operational High Risk Practices."

The critical items are:

Time/temperature
Cooking, cooling, reheating, and holding
Preparation ahead of time
Cross-contamination
Personal hygiene

REVIEW OPERATIONAL PRACTICES THAT POSE HIGH RISK

* VOLUME THAWING—MEAT AND POULTRY OR PHF

* VOLUME PREPARATION—POTENTIALLY HAZARDOUS FOOD

* PREPARATION FROM SCRATCH

* SIMULTANEOUS PREPARATION OF RAW AND UNCOOKED FOODS

* PREPARATION A DAY IN ADVANCE

* VOLUME COOLING

* USE OF LEFTOVERS

* LACK OF SANITATION AWARENESS

* HIGH DEGREE OF FOOD HANDLING AND CONTACT

* EMPLOYMENT OF VOLUNTEERS AND HIGH STAFF TURNOVER

* INFREQUENT TRAINING

* QUESTIONABLE ADEQUACY OF HOT AND COLD HOLDING EQUIPMENT

* TRANSPORT OF FOOD

* SERVING FRAIL, SICK, ELDERLY, OR IMMUNE-COMPROMISED INDIVIDUALS

The critical items most likely to be implicated with operational practices that pose high risk are as follows:

Volume Thawing
    Time/temperature
    Cooking, cooling, reheating, holding

Volume Preparation
    Time/temperature
    Cooking, cooling, reheating, holding
    Preparation ahead of time
    Cross-contamination
    Personal hygiene

Preparation from Scratch
    Cooking, cooling, reheating, holding
    Preparation ahead of time
    Cross-contamination
    Personal hygiene

Simultaneous Preparation of Raw and Uncooked Foods
    Cross-contamination
    Personal hygiene

Preparation a Day in Advance
    Time/temperature
    Cooking, cooling, reheating, holding
    Preparation ahead of time

Volume Cooling
    Time/temperature
    Cooking, cooling, reheating, holding

Use of Leftovers
    Time/temperature
    Cooking, cooling, reheating, holding
    Preparation ahead of time

Lack of Sanitation Awareness
    Cross-contamination
    Personal hygiene

High Levels of Food Handling and Contact
    Cross-contamination
    Personal hygiene

Critical Items and Food Flow

The HACCP food safety system treats the storage, preparation, and service of foods as a continuous process. First you learned which foods (PHFs) are most often linked to food-borne illness. Now, by focusing on critical items, you can zero in on the problems and practices that are the leading factors in foodborne illness. Thus, you will be able to quickly identify any potential problems. Once you identify the problems, you will be able to take the necessary steps to correct them.

Using the HACCP steps, you start by following a potentially hazardous food through a typical flow process and identifying the critical items in that sequence. For a foodservice establishment, a typical flow would include:

- Menu Planning
- Recipe Development
- Purchasing and Receiving
- Storage
- Reconstitution
- Thawing
- Handling and Preparation
- Cooking
- Hot Holding
- Cooling and Cold Holding
- Reheating
- Serving or Repacking
- Cleaning
- Hygiene of Workers

Review the "Processing Flow Chart" on page 42. Identify the critical items to be controlled at each step.

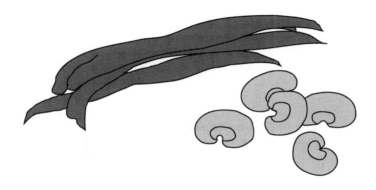

Processing Flow Chart

Turkey

Source		Delivery		Storage
Contamination		Time/Temperature		Time/Temperature Cross-Contamination

Chilling		Boning		Cooking
Time/Temperature Cooling Cross-Contamination Personal Hygiene		Time/Temperature Holding Cross-Contamination Personal Hygiene		Time/Temperature Cooking, Holding Preparation Ahead of Time

Reheating		Holding		Serving
Time/Temperature Reheating Cross-Contamination Personal Hygiene		Time/Temperature Hot Holding		Cross-Contamination Personal Hygiene

Food Flow Considerations

Menu Planning

- Do certain menu items present particular food safety risks—raw oysters or fresh whole eggs, for example?
- If you are using these food items, what safety precautions and staff training must be built into your food safety system?
- What is the makeup of your customer base that requires special precautions: children, people who are elderly, pregnant, or immune compromised?
- Do you have the proper equipment to accommodate your menu and volume?
- What is the employee skill level and training required for the menu?
- Do you have a high employee turnover rate that may require menu considerations?

Recipe Development

- Identify recipe ingredients that require special attention.
- For food safety, consider substitute ingredients, such as pasteurized eggs instead of fresh whole eggs.
- Write recipes to include safe food handling procedures and temperatures.
- Write procedural standards for advance and/or large volume preparation, cooling, and handling.
- Write procedural standards for off-premise food handling, transporting, holding, and reheating.

Purchasing

- Purchase from approved sources.
- Are delivery trucks refrigerated?
- What is the handling practice during transportation and receiving delivery?
- Can deliveries be scheduled so that they can be properly inspected and stored?
- Are products consistent?
- Are products safety packaged?

Receiving

- Is packaging clean and intact?
- Is there any off-odor or slickness on the product?
- Are employees trained in receiving/storage procedures?
- Check product temperatures; refrigerate PHFs immediately. Do not puncture sealed packaging; take temperature by placing thermometer between two packages or underneath packaging.
- Check for cross-contamination of foods, such as poultry juices dripping on other products.
- Are shellfish tags for mussels, oysters, and clams dated, logged, and saved for 90 days?
- Follow established company policies and procedures for rejecting products.

Storage

- Do storage practices prevent cross-contamination?

    Store raw meats on lower shelf.

    Raw vegetables or uncooked menu items should be stored above raw potentially hazardous foods.

- Label, date, and use FIFO (first in, first out) rotation.
- Refrigerate meat and other PHFs at 40°F or below.
- Do employee personal hygiene practices prevent contamination of food items?

Preparing/Cooking

- Wash hands.
- Clean and sanitize utensils, cutting boards, and knives.
- Preplan product needs and thaw foods under refrigeration.
- What are the batch preparation procedures?

    Work on small units of food at one time, then refrigerate.

    Are ingredients pre-chilled for salad preparation?

- Wash vegetables in sanitized sink.
- Cook all PHF menu items to recommended minimum temperatures.
- Verify food temperatures with a calibrated thermometer.
- Use proper tasting procedures using a clean spoon or a clean sauce dish each time the product is tasted.

Serving/Holding

- Use clean, sanitized equipment to transfer food and hold products.
- Use clean and sanitized utensils.
- Set up stations and product handling process to prevent cross-contamination.
- Monitor employee personal hygiene practices.
- Hold hot food items at 140°F or higher.
- Hold cold food items at 40°F or lower.
- Verify food temperatures with a thermometer.
- Keep food covered.

Cooling

- Rapidly cool foods from 140°F to 70°F in less than two hours and from 70°F to 40°F in four hours or less.

    Use ice baths.

    Use shallow pans with less than 3 inches of product.

    Divide into smaller units.

    Withhold water; add ice as part of liquid at end to cool down products and utensils.

- Prevent cross-contamination when using ice bath.
- Write cool-down procedures into recipes.

- Verify final temperature with a calibrated thermometer.
- Use clean, sanitized pans.
- Cover immediately after cooling to 40°F.
- Store on top or upper shelves of refrigerator.
- Label food item; include date and time.
- Do not stack pans; stacked pans cool as one large mass.

Reheating

- Heat rapidly to 165°F within two hours.
- Determine which equipment or methods work best for reheating.
- Verify final temperature with a calibrated thermometer.
- Maintain temperature of 140°F; verify with a calibrated thermometer.
- Never mix new product into old product.
- Do not reheat or serve leftover food more than once; a reheated product passes through the temperature danger zone three times.

It is possible
to fail in many
ways . . .
while to succeed
is possible only
in one way.
                    –Aristotle

Control Point (CP)

Any point, step, or procedure at which biological, physical, or chemical factors can be controlled.

Critical Control Point (CCP)

A point, step, or procedure in the product-handling process where controls can be applied and a food safety hazard can be prevented, eliminated, or reduced to acceptable levels.

A critical control point is the "kill" step in which the bacteria are killed by cooking or the "control" step that prevents or slows their growth, such as proper chilled storage or hot holding.

Critical Control Point Guidelines*

- At this step of preparation can:
  - ○ Food become contaminated?
  - ○ Contaminants increase?
  - ○ Contaminants survive?

- Can this hazard be prevented through corrective action(s)?

- Can this hazard be prevented, eliminated, or reduced by steps taken later in the preparation process?

- Can you monitor the critical control point (CCP)?

- How will you measure the CCP?

- Can you document the CCP?

*Adapted with permission from Managing a Food Safety System seminar. Copyright © 1992 by the Education Foundation of the National Restaurant Association.

Food Flow Critical Control Points

At first there is a tendency to identify all control points as critical control points (CCPs). When determining the CCPs in a product's flow, select the monitoring points at which a hazard can be prevented, eliminated, or reduced. These should become the checkpoints in your system, recipe, or procedures. The following are examples of CCPs in a food flow:

Receiving

- Meats must be received at 40°F or lower.
- Frozen foods must be received at 0°F.
- Fresh fish must be received at 40°F or lower.
- Fresh fish require a Certificate of Conformance from the supplier.

Cooking

- Poultry must be cooked to165°F or higher.
- Gound meats must be cooked to 155°F or higher (except ground poultry, which must be cooked to 165°F).
- Pork must be cooked to 155°F or higher.

Hot-Holding

- Products must be held at 140°F or higher.

Cooling

- Cool from 140°F to 45°F in four hours (current code).
- Cool from 140°F to 70°F in two hours or less and from 70°F to 40°F in four hours or less (proposed new code).

Cold Holding

- Product must be held at 40°F or lower.

Reheating

- Reheat to 165°F within two hours.

Serving ("Standard Operating Procedure" controls)

- Monitor employees' hygiene.
- Monitor for cross-contamination.

FOOD FLOW CHART
Hamburger on Bun

RECEIVING AND STORAGE

CP | Ground Beef patties received at 0°F or lower.

(Control point [CP] is not a CCP, as cooking to proper temperature in the next step will prevent a hazard).

COOKING

CCP | Cook to 155°F or above until meat is no longer pink and juices run clear (kills *E. coli*).

Check batches with clean, sanitized thermocouple.

(CCP—Proper cooking temperature is required to prevent a hazard.)

HOLDING

CCP | Hold at 140°F or above.

(CCP—Proper holding temperature prevents the growth of organisms.)

Refried Rice

STORING

Store rice in dry area. (Grains, rice, and pasta do not become hazardous until water is added.)

COOKING

CCP

Combine with water.
Bring to boil, simmer until tender. (Heat to 165°F or above.)

COOLING

CCP

Place rice loosely in shallow pans with product depth of 2 inches or less.
Rapidly cool to 40°F within six hours or less (prevents rapid bacterial growth and spore formation when cooled quickly).

REHEATING

CCP

Reheat to 165°F or higher within two hours or less (prevents rapid bacterial growth and spore formation when reheated rapidly).
Reheat only one time. (Reheated products have passed through the temperature danger zone three times.)

HOLDING

CCP

Hold for service at 140°F or above (prevents bacterial growth).

Fresh Grilled Salmon on Baby Greens

PURCHASING

CCP Purchase fish from approved, certified source (certifies seafood is from safe sources; cooking does not destroy toxins).

RECEIVING AND STORAGE

Receive and store fresh fish at 40°F or below.

PREPARATION AND HOLDING

Wash all salad ingredients.
Assemble salads following recipe food safety instructions.
CCP Hold all salad ingredients at 40°F or below (prevents rapid bacterial growth in items that will not receive further cooking).

COOKING

CCP Grill salmon until fillet reaches an internal temperature of 145°F or higher (kills parasites & bacteria).
Wash hands after handling raw fish.

SERVING

Arrange grilled salmon on plate with greens.
Complete garnishing of salad.
Serve immediately.

Broccoli and Cheese Quiche

RECEIVING AND STORAGE

Receive and store pasteurized eggs at 40°F or below.
Purchase only pasteurized milk products.

Refrigerate cheese and vegetables at 40°F or below after receiving.

PREPARATION

Wash broccoli and other vegetables.
Prepare vegetables with clean, sanitized utensils.

COOKING

Saute vegetables lightly.

Stir in grated cheese, cream, and pasteurized eggs.

CCP Bake until quiche reaches internal temperature of 145°F or higher. (Pasteurized eggs need to be cooked to only 145°F to kill bacteria.)

HOLDING

CCP Hold at 140°F or above. (All egg products need to be held at 140°F to control bacteria growth.)

FOOD FLOW CHART
Creamed Chicken Soup

RECEIVING AND STORAGE

Receive and store PHFs at 40°F or below.

Store raw PHFs on lower shelf of refrigerator.

Use pasteurized dairy products.

PREPARATION

Use cleaned and sanitized utensils.

Refrigerate batches as completed.

Wash hands before handling vegetables.

Use clean and sanitized sink for washing and preparation of vegetables.

COOKING

Cook to minimum of 165°F.

HOLDING

Hold at 140°F or above.

COOLING

Cool rapidly from 140°F to 70°F in two hours or less and from 70°F to 40°F in four hours or less.

REHEATING

Reheat rapidly to 165°F within two hours.

Review the Food Flow Chart for Creamed Chicken Soup and mark the CCPs. A CCP is the step which if not performed correctly, may result in a hazard that would not be prevented by a following step. The correct CCPs are shown on page 54.

FOOD FLOW CHART
Creamed Chicken Soup

RECEIVING AND STORAGE

Receive and store PHFs at 40°F or below.

Store raw PHFs on lower shelf of refrigerator.

Use pasteurized dairy products.

PREPARATION

Use cleaned and sanitized utensils.

Refrigerate batches as completed.

Wash hands before handling vegetables.

Use clean and sanitized sink for washing and preparation of vegetables.

COOKING

CCP Cook to minimum of 165°F (cooking kills bacteria).

HOLDING

CCP Hold at 140°F or above (prevents growth of bacteria).

COOLING

CCP Cool rapidly from 140°F to 70°F in two hours or less and from 70°F to 40°F in four hours or less (prevents the rapid growth and spore formation of bacteria such as *C. perfringens*, which likes the environment of products such as stews, soups, and sauces).

REHEATING

CCP Reheat rapidly to 165°F within two hours. (Prevents the rapid growth of *C. perfringens* and its ability to form spores that can survive the heating process. When this bacteria is ingested in large amounts it forms toxins in the intestines, making people ill.)

Critical Food Temperatures*

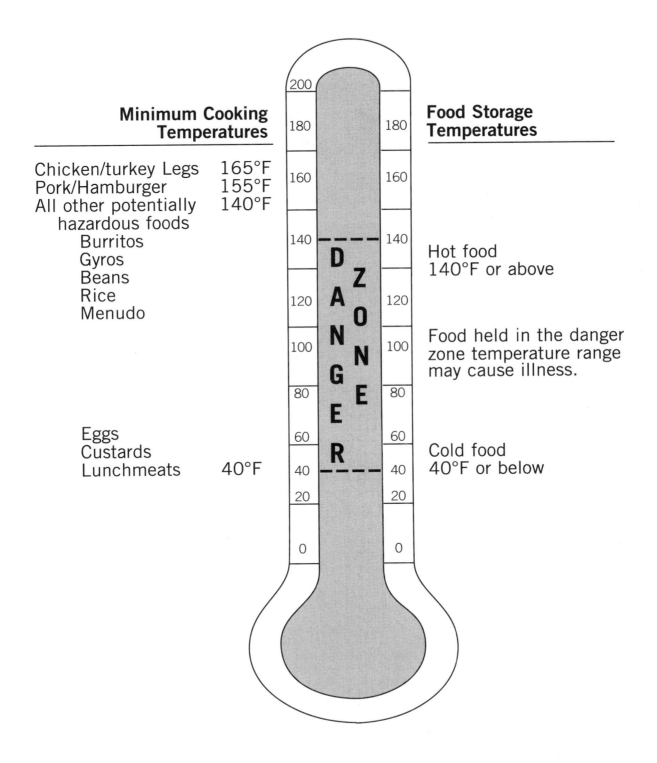

Minimum Cooking Temperatures		Food Storage Temperatures
Chicken/turkey Legs	165°F	
Pork/Hamburger	155°F	
All other potentially	140°F	
hazardous foods		
Burritos		Hot food
Gyros		140°F or above
Beans		
Rice		
Menudo		Food held in the danger zone temperature range may cause illness.
Eggs		
Custards		Cold food
Lunchmeats	40°F	40°F or below

Thermometer markings (left and right): 200, 180, 160, 140, 120, 100, 80, 60, 40, 20, 0

DANGER ZONE (140°F to 40°F)

Colorado Department of Health and Denver Department of Public Health/Consumer Protection.

Temperaturas Criticas de Alimentos*

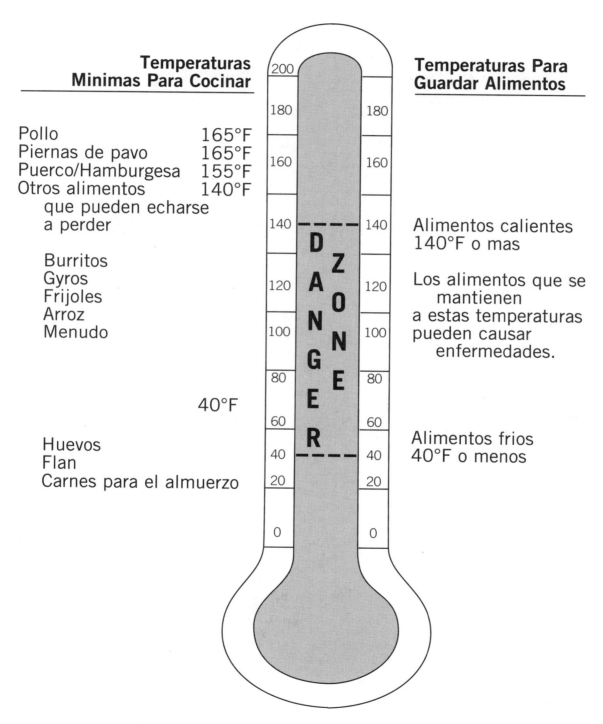

Temperaturas Minimas Para Cocinar

Pollo	165°F
Piernas de pavo	165°F
Puerco/Hamburgesa	155°F
Otros alimentos	140°F

que pueden echarse a perder

Burritos
Gyros
Frijoles
Arroz
Menudo

40°F

Huevos
Flan
Carnes para el almuerzo

Temperaturas Para Guardar Alimentos

Alimentos calientes 140°F o mas

Los alimentos que se mantienen a estas temperaturas pueden causar enfermedades.

Alimentos frios 40°F o menos

Thermometer markings: 200, 180, 160, 140, 120, 100, 80, 60, 40, 20, 0 — DANGER ZONE

*Colorado Department of Health and Denver Department of Public Health/Consumer Protection.

Employee Hygiene and Health

Good personal hygiene is essential for food safety and to protect against foodborne illness. Infected persons and poor personal hygiene account for about twenty-five percent (25%) of foodborne illness outbreaks.

Contaminated hands transmit bacteria and viruses from the body and from feces to food. Managers must adopt a policy that employees with diarrhea and flulike symptoms do not work around food.

Employees may also feel fine and still be infective. Such is the case when a worker is infected with *Hepatitis A*. A person can be shedding the virus for about 20 days before he or she shows symptoms of the illness. If proper hand-washing practices have been established, they may be effective in controlling the spread of illness.

Sooner or later the idea here put forth will conquer the world, for with inexorable logic it carries with it both the heart and the head.

–Albert Schweitzer

Food Safety Standards for Employees

The following are guidelines that must be followed by all foodservice employees.

ILLNESS

Anyone who is sick should not work with food. Inform your supervisor if you have a severe cold or diarrhea.

CUTS, ABRASIONS, AND BURNS

Wounds should be bandaged antiseptically. Cover bandages with waterproof protection such as rubber gloves or finger cots. Inform supervisor of all wounds.

HAND WASHING

1. Thoroughly wash hands and exposed portion of your arms with soap and warm water.

   - Before starting work.
   - During work as often as needed to keep clean.
   - After break time.
   - After touching anything that can be a source of contamination:
       Telephone, money, soiled linens
       Raw foods, meats, shell eggs, fresh produce
       Dirty dishes, equipment, utensils, or trash
   - After using cleaners or chemicals.
   - After performing personal needs, such as smoking, eating, drinking, sneezing, coughing, or using the toilet.
   - After picking up items off the floor.

2. Follow proper steps in hand washing.

   - Use soap and warm (105°F) running water.
   - Rub your hands vigorously for 20 seconds.
   - Wash all surfaces, including:
       Backs of hands
       Wrists
       Between fingers
       Under fingernails
   - Use nail brush around and under fingernails.
   - Rinse well under running water.
   - Dry hands with paper towel.
   - Turn off the water and open door knobs using a paper towel instead of bare hands.

3. Make sure all hand washing stations and rest rooms are well stocked with soap and towels.

PLASTIC GLOVES

If plastic gloves are utilized, wear them over thoroughly washed hands. The loose-fitting style is recommended. Change gloves often and under the same circumstances as you would wash your hands.

FINGERNAILS

Keep fingernails clean and trimmed. False fingernails and nail polish should not be worn as they can chip or break off into the food.

JEWELRY

Jewelry is not to be worn, as it gets dirty, can get lost in food, or can even cause injury when caught by a hot or sharp object or equipment.

UNIFORMS

Uniforms are to be clean, changed daily, and worn in designated areas. Wear clean clothes to work and change only in locker rooms. Uniforms should not be worn to and from work.

APRONS

Wear a clean apron. An apron should not be used as a hand towel. Follow hand washing procedures after touching or wiping your hands on an apron. Remove your apron when leaving the food preparation area.

HAIR RESTRAINTS

Hats and hairnets are considered proper hair restraints. Hair restraints are required to keep hair and its contaminants out of food. After touching hair or face, follow proper hand washing procedures.

SMOKING, EATING, GUM CHEWING, CHEWING TOBACCO, TOOTHPICKS

Smoke only in designated areas, and eat only in the employee dining room. Do not chew gum while working. Follow proper hand washing procedures after smoking, eating, drinking, or chewing. This is important in preventing mouth to hand contamination.

GROOMING

Bathe daily, use a deodorant, change into clean clothes daily. Wear proper work shoes and keep them cleaned.

Controlling Infectious Disease

A wide range of communicable diseases and infections may be transmitted by infected food employees. Proper management begins with employing healthy employees and instituting a system to identify employees who present a risk of transmitting foodborne pathogens to food or to other employees.

It is the responsibility of management to convey to applicants and employees the importance of notifying the person in charge of any changes in an employee's health status. Once notified, the person in charge must take action to prevent the likelihood of the transmission of foodborne illness. Food employees or applicants are required to report if they:

1. Are diagnosed with an illness caused by:
    Salmonella typhi
    Shigella spp.
    Escherichia coli 0157:H7
    Hepatitis A virus infection

2. Have symptoms of intestinal illness

3. Have boils or infected wounds

4. Have a high risk of becoming ill:
    Prepared or consumed food that caused disease outbreak
    Live with an ill person
    Live with a person involved in a disease outbreak
    Traveled outside the United States in the last 50 days.

The following work status guidelines must be followed when food handlers are diagnosed with these diseases.

FOOD HANDLER WORK RESTRICTIONS*

Disease	Work Status	Duration of Work Restriction / Comments
Abscess, boils, etc.	Relieve from direct contact and food handling.	Until drainage stops and lesion has healed or employee has negative culture.
AIDS (acquired immune deficiency syndrome) or ARC (AIDS-related complex)	May work (per CDC Guidelines).	Employee will be counseled, educated.
Diarrhea		
a. Acute stage (etiology unknown)	Relieve from direct food handling.	Until symptoms resolve and infection with salmonella, shigella, or campylobacter is ruled out.
b Campylobacter	Relieve from direct food handling.	Until symptoms resolve or after appropriate antibiotic therapy for 48 hours.
c. Salmonella	Relieve from direct food handling.	Until stool is free of the infecting organism in two consecutive cultures, not less than 24 hours apart.
d. Shigella	Relieve from direct food handling.	Until stool is free of the infecting organism in two consecutive cultures, not less than 24 hours apart.
Hepatitis A	Relieve from direct food handling.	Until seven days after onset of jaundice. Must bring note from physician upon return.
Staphylococcus aureus (skin lesions)	Relieve from direct food handling.	Until lesions have resolved and the employee has a negative culture.

*Centers for Disease Control.

Why Proper Hand Washing Is Essential*

Emphasis on personal hygiene for foodservice personnel is fundamental for food protection and sanitation practice. Hand washing is probably the most important aspect of personal cleanliness. Unclean hands can easily transmit microbial agents to food products.

Management has the responsibility to provide (with adequate maintenance and supplies) hand-washing sinks, hot water, soaps, and paper towels (or air dryers) in kitchen and rest room areas, and to encourage employees to use the facilities throughout the workday as necessary. Sinks used to prepare foods *must not* be used to wash hands.

Foodservice personnel have the responsibility to practice good personal hygiene with special attention to washing their hands when beginning work, frequently (and effectively) during the work period, and *every* time after having performed the activities shown in the "WASH HANDS" list on page 66.

Many tasks in the kitchen involve use of the food worker's hands. *Frequent hand washing is a significant factor in the prevention of illness.* Hand washing cannot be overemphasized.

Good hand habits are essential in foodservice to provide clean, safe food to the customer. Follow the hand-washing procedures outlined on pages 66–68 to ensure a thorough cleaning of the hands.

As adequate hand washing is the cornerstone of aseptic practice in medical facilities, it is undoubtedly a most important factor in preventing the transmission of disease organisms in a food establishment.

*University of Massachusetts/Amherst, *Environmental Health and Safety,* May, 1989.

DROWN A GERM

WASH
YOUR HANDS!

WASH HANDS

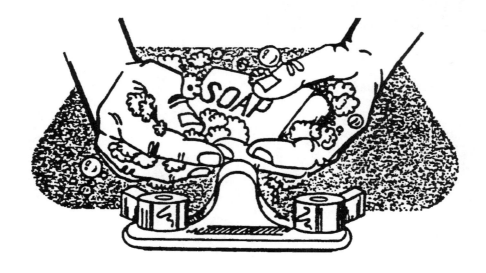

○ BEFORE STARTING WORK

○ AFTER USING THE BATHROOM

○ AFTER BREAK TIME, SMOKING, EATING, OR DRINKING

○ AFTER CHEWING GUM OR USING TOOTHPICKS

○ AFTER COUGHING, SNEEZING, BLOWING OR TOUCHING YOUR NOSE

○ AFTER TOUCHING HEAD, HAIR, MOUTH, WOUNDS, OR SORES

○ AFTER TOUCHING RAW POULTRY, MEATS, OR FISH

○ AFTER TOUCHING DIRTY DISHES, EQUIPMENT, AND UTENSILS

○ AFTER TOUCHING TRASH, FLOORS, SOILED LINENS, ETC.

○ AFTER USING CLEANERS OR CHEMICALS

○ DURING FOOD PREPARATION AS NECESSARY

LAVESE LAS MANOS

○ ANTES DE EMPEZAR A TRABAJAR

○ DESPUES DE IR AL BAÑO

○ DESPUES DEL DESCANSO, FUMAR, COMER, O BEBER

○ DESPUES DE TOSER, ESTORNUDAR

○ DESPUES DE MASTICAR CHICLE O DE USAR PICADIENTES

○ DESPUES DE TOCARSE LA CABEZA, EL PELO, LA BOCA, HERI-
DAS, O ULCERAS

○ DESPUES DE TOCAR CARNES CRUDAS, CARNES, O MARISCOS

○ DESPUES DE TOCAR LOZA SUCIA, EL EQUIPO, O UTENSILIOS

○ DESPUES DE TOCAR BASURA, PISOS, MANTELERIA SUCIA, ETC.

○ DESPUES DE USAR LIMPIADORES O QUIMICAS

○ CUANDO SEA NECESARIO DURANTE LA PREPARACION DE
ALIMENTOS

Wash Your Hands After*

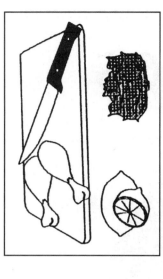

Handling raw food — particularly meat, poultry, and foods served uncooked.

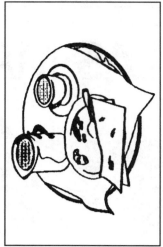

Handling dirty dishes and before handling clean dishes.

Going to the restroom.

Smoking, eating, or drinking.

Touching or scratching any area of the body (ears, mouth, nose, hair, etc.).

Touching any soiled object or surface, soiled clothing, etc.

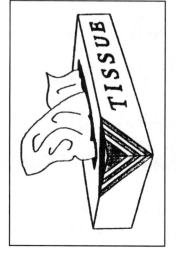

Using a handkerchief or tissue.

Cleaning, taking out the trash, or putting away supplies.

*Denver Department of Health Consumer Protection Division (303) 436-7330.

Proper Use of Disposable Gloves

Nothing can replace good sanitary practices in food preparation—whether at home, at a restaurant, at an outdoor event, in a warehouse, or when catering an event. These practices include the following:

- *Wash hands before preparing foods*, after toileting, and any time hands may be contaminated by urine, feces, saliva, or mucus.
- *Minimize manual contact with foods* through the use of gloves, tongs, or other suitable utensils.

The use of gloves is recommended when handling fresh products, that will not be cooked, and products that have been cooked and will receive no further heat treatment.

When gloves are used, extra care must be taken to prevent a false sense of security. Improperly used gloves carry a high risk for cross-contamination, because workers may not be aware that gloves are contaminated and should be changed.

- *Wash hands before using gloves.* This prevents contamination of the gloves by the hands.
- *Wash hands after using gloves.* The warm, moist environment provided by the gloves allows bacteria to grow and multiply on the hands.
- *Gloves should be discarded* when they become soiled or when the product being handled changes from raw to finished. The general rule is: *Gloves should be changed according to the same rules that dictate when hand washing should occur.*
- *Discard gloves when leaving the work area,* even to get supplies, because the gloves are considered to be contaminated by touching door handles and equipment. When returning to the work area, wash hands and use a fresh pair of gloves.
- *It is more economical to discard soiled gloves* according to the recommended guidelines, than it is to treat customers and employees who become ill owing to poor practices.

HAND-WASHING
Procedure—Single Wash Method

1. Use SOAP and WARM RUNNING WATER.
2. RUB your hands vigorously for 20 seconds.
3. WASH ALL SURFACES, including:
   backs of hands
   wrists
   between fingers
   under fingernails
4. WASH and BRUSH EACH FINGER.
   Use a nail brush.
5. RINSE well under running water.
6. DRY hands with a paper towel.
7. Turn off the water using a PAPER TOWEL instead of bare hands.
   OPEN rest room doors with PAPER TOWEL, then DISCARD paper towel.

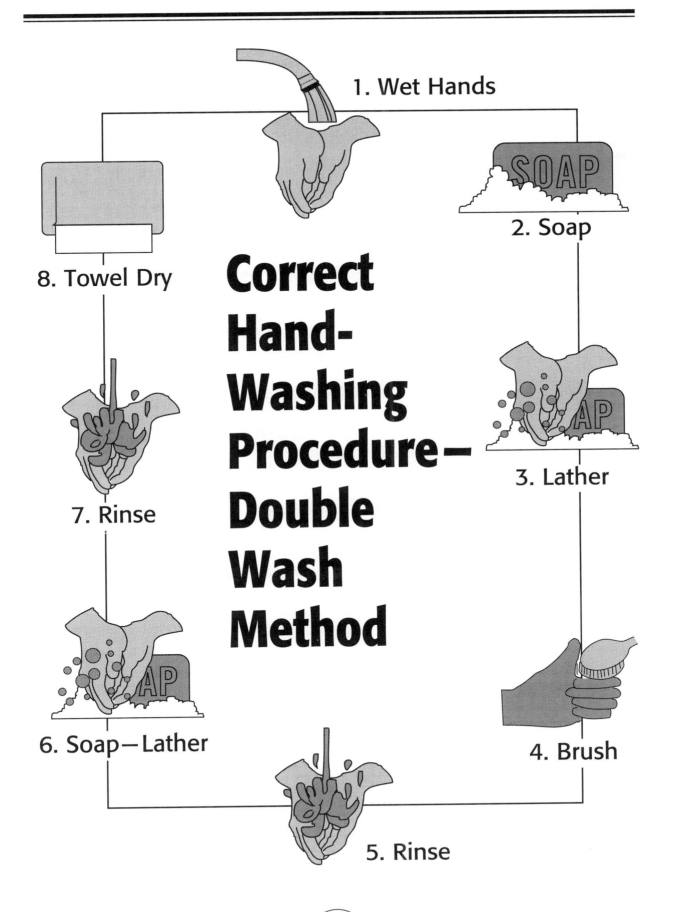

1. Wet Hands

2. Soap

3. Lather

4. Brush

5. Rinse

6. Soap—Lather

7. Rinse

8. Towel Dry

Correct Hand-Washing Procedure—Double Wash Method

Cross–Contamination

Contamination is the unintended presence of harmful substances or organisms in food.

Cross-contamination is the transfer of harmful bacteria from one food to another by means of a nonfood surface, such as utensils, equipment, or human hands.

Cross-contamination can be controlled by adhering to safe food handling practices such as:

- Following proper hand-washing procedures.
- Restrict from working any employees with flulike symptoms.
- Use utensils or disposable gloves when handling food that will no longer be cooked.
- Follow rules for proper use and disposal of gloves.
- Avoid preparing raw and cooked foods in same work area.
- Store raw and cooked foods in separate areas.
- Store raw meats below cooked foods or foods that will no longer be cooked.
- Follow rules for proper use and concentration of sanitizers in work areas and on wiping cloths.
- Follow rules for proper washing and sanitizing of equipment, utensils, knives, and dishes.

Special attention to prevent cross-contamination is required in some work areas where the foodservice worker is required to handle many different products at once. The grill or broiler station and prep stations are examples. In these work areas the chef or cook is handling raw meat products, raw eggs, and foods that will no longer be cooked, such as breads and buns, garnishes, cheeses, and other cold condiments.

On page 73 is a flow chart of a typical grill line. Follow the arrows; notice the possibility for cross-contamination. Which foods require special handling to prevent cross-contamination? What is your recommendation? See answers on pages 74 and 75.

Grill Line Flow Chart

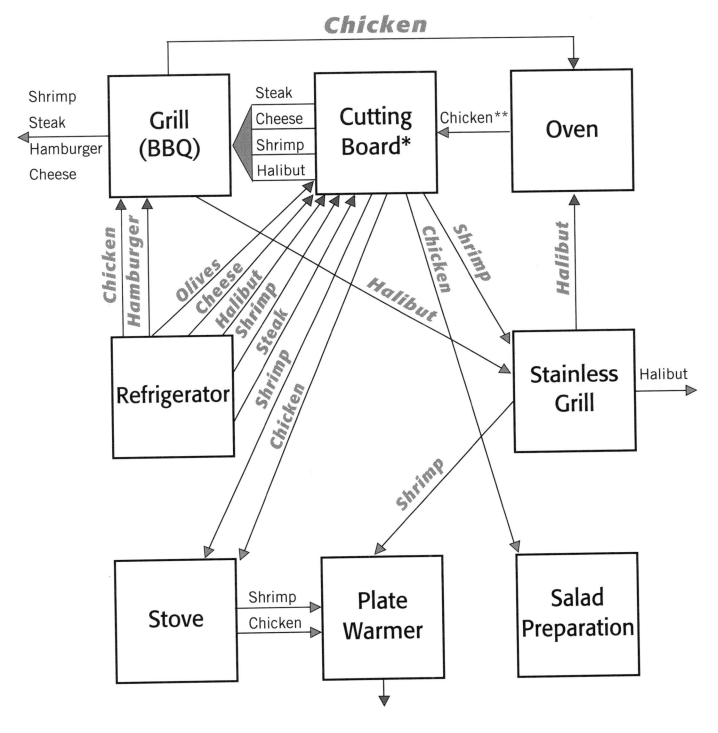

*Cooked chicken, chopped and held in hot kitchen area until ordered for various dishes

**Chicken cooked to a range of 140–168°F

Answers are on pages 74 and 75.

Cross-Contamination by Use of Cutting Board

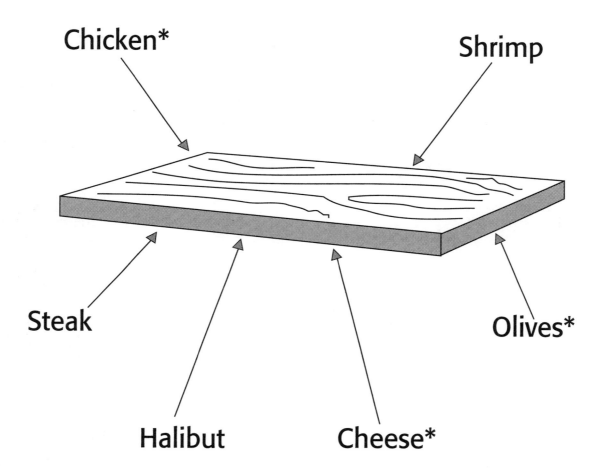

Chicken*

Shrimp

Steak

Olives*

Halibut

Cheese*

*Foods served RAW or WITHOUT FURTHER COOKING after coming into contact with contaminated cutting board

Cross-Contamination by Hands

Raw

Chicken

Halibut

Shrimp

Steak

Cheese

Hamburger

Cooked

Chicken

Halibut

Shrimp

Steak

Cheese

Hamburger

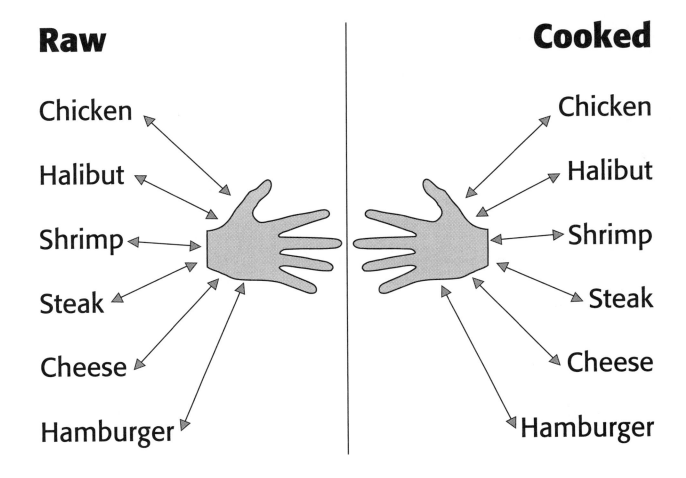

Chef alternately handled various raw and cooked food, which allowed cooked food to be cross-contaminated with raw meat and seafood.

Preventing Cross-Contamination —
Grill or Prep-at-Service Stations

Cross-contamination can be prevented by properly handling and storing foods during production.

- Store each type of raw meat in separate containers. For example, chicken and fish require different final cooking temperatures. If stored together, both products require a higher final cooking temperature of 165°F. If stored separately, fish needs to be cooked only to 145°F.

Solutions

Eliminate cross-contamination opportunities by frequently cleaning and sanitizing equipment, utensils, and work surfaces during production.

- The use of sanitizing solutions, hand washing, and sound personal hygiene habits must be followed during all steps of line prep-at-service. Ensure that sanitizer buckets are provided and used.
- A set of tongs should be designated for each raw food and stored with handles out of the food for safe use.
- Pay special attention to the plating at the last step of prep-at-service when toppings are placed on plated foods. Tongs are to be used at all times with items such as shredded cheese, garnishes, and foods that will not receive additional cooking.
- When stocking line drawers or refrigerators keep in mind that storing food items behind one another creates a cross-contamination concern because of dripping juices. It is a good standard of practice to keep raw food in the front of the lower drawers.
- Ensure that all employees can identify raw and cooked foods and that they understand where these foods should be stored. Verify storage practices during each meal check.
- Cooked and raw foods should never be stored in the same container. Use food containers to hold food and change containers each time the food containers are to be refilled to prevent cross-contamination.
- Build in operational systems that will allow foods (i.e., raw chicken and veal) to be handled at the end of the production day to minimize the opportunity for cross-contamination.

Safe Handling of Cold Foods —
Preventing Cross-Contamination

When working with cold foods or foods that will not be further heat treated, follow these general guidelines:

- Wash hands with soap and warm water and dry with single use towel before starting prep.
- Wash all fresh fruit and vegetables before using to remove germs, soil, and chemicals.
- Prep sinks are not to be used as hand sinks. Such use contaminates the sinks.

- Prep sinks should not be used for meat, poultry, and fish items; however, if there is no other option, wash and sanitize after each use and before preparing items.
- Raw vegetable preparation must be done in a separate area from raw meat preparation.
- Use separate and sanitized cutting boards for raw and uncooked items.
- Colored cutting boards help to identify boards used for raw products and prevents cross-contamination.
- Pre-chilling salad ingredients can help control hazards. Store tuna and mayonnaise in the refrigerator and cool down cooked chicken, pasta, potatoes, and other ingredients before mixing. To quickly chill salad ingredients:
  Cut into smaller pieces and place in thin layer on pan in coldest part of refrigerator.
  Place ice on pasta or surround ingredients with bagged ice.
- When mixing ingredients avoid hand contact. Use clean, sanitized utensils or a fresh pair of disposable gloves.
- Transfer ingredients to clean, sanitized dishes or clean storage containers.
- Store and display at 40°F or colder.
- Protect cooked meats and ready-to-eat raw fruits and vegetables from raw meat juices during storage.
- After coughing, sneezing, or touching face or hair, wash your hands before handling food or utensils.

Sanitizing Procedures

Sanitizing is the reduction of microorganisms to safe levels. Before sanitizing can be achieved, soil and food particles must be removed from the surface of equipment and utensils. After washing to remove the food particles and soil, the surface must be rinsed. Rinsing removes the detergent and any remaining loose particles. Detergents must be rinsed off as they reduce the effectiveness of the sanitizer solution. Sanitizing is not a substitute for cleaning.

There are two ways to sanitize:

1. Heat—Rinse temperatures need to be greater than 170°F.

2. Chemical—Sanitizer solutions are most effective in the temperature range of 75°F–120°F.

The correct concentrations of various sanitizers are shown in the following table.

Chlorine	Iodine	Quaternary Ammonia
1 teaspoon per gallon 50–100 ppm*	1 ounce (2 tablespoons) per 5 gallons = 25 ppm 12.5–25 ppm*	Quaternary concentrations vary. You need to test the quaternary ratio to water for your brand. 100–200 ppm*

*Use test strips to determine the proper strength of sanitizer. Each type of sanitizer requires its own test strip. More sanitizer is not better; too much is toxic and can cause chemical poisoning. Test strips are available from your local supplier or chemical company. If you are not familiar with how to use test strips, ask your chemical representative for a demonstration.

The proper use of sanitizers is essential to prevent cross-contamination. Are sanitizers being used in your operation? Are there buckets of sanitizer solution at the work stations? Does the solution constantly look clean—a sign that it is not being used!

To sanitize a surface properly:

1. First wash with warm soapy water.

2. Then rinse with clean water.

3. Finally, wipe with sanitizer solution. *Do not rinse or towel dry.*

CORRECT TABLE AND KITCHENWARE WASHING BY HAND*

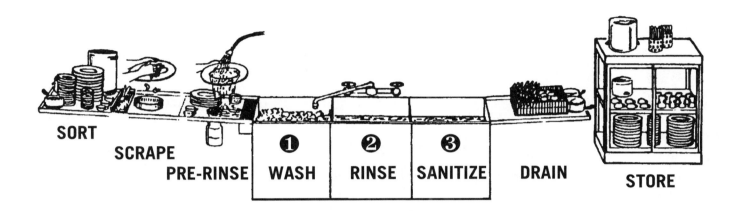

SORT

SCRAPE

PRE-RINSE ❶ **WASH** ❷ **RINSE** ❸ **SANITIZE** **DRAIN**

STORE

❶ WASH

WITH A GOOD
DETERGENT
IN WATER AT NOT
LESS THAN 110°F

❷ RINSE

IN CLEAN
HOT WATER

❸ SANITIZE

IN WATER AT NOT
LESS THAN 170°F
(OR) ADEQUATE
CHEMICAL SOLUTION
FOR 1 MINUTE

	(ppm) min-max
Chlorine	50–100
Iodine	12.5–25
Quat. Amm.	100–200

*Denver Environmental Health Services, Consumer Protection/Food Safety.

EL MODO CORRECTO DE LAVAR LOS PLATOS A MANO*

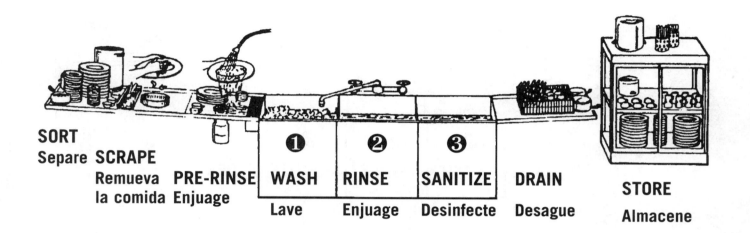

SORT
Separe **SCRAPE**
Remueva **PRE-RINSE**
la comida Enjuage

❶ **WASH**
Lave

❷ **RINSE**
Enjuage

❸ **SANITIZE**
Desinfecte

DRAIN
Desague

STORE
Almacene

❶ LAVE	❷ ENJUAGE	❸ DESINFECTE
CON UN BUEN DETERGENTE A UNA TEMPERATURA MINIMA DE 110°F	CON AGUA LIMPIA Y CALIENTE	CON AGUA A UNA TEMPERATURA MINIMA DE 170°F (O) CON UNA SOLUCION QUIMICA POR UN MINUTO

(ppm) minimo-maximo
Chlorine   50–100
Iodine   12.5–25
Quat. Amm.   100–200

Denver Environmental Health Services, Consumer Protection/Food Safety.

Sanitation Procedure—Slicer

Clean and Sanitize

- After each use period
- At least once every 4 hours

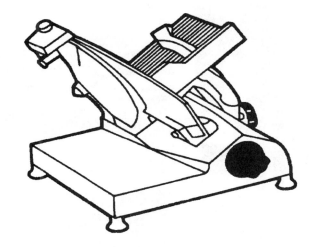

1. Unplug slicer. Set the blade control to 0.
2. Disassemble parts.
3. Wash all removable parts with hot soapy solution.
4. Rinse with clear water.
5. Sanitize by dipping into an approved sanitizer solution.
6. Let air dry.
7. Use a long-handled brush or thick cloth to clean stationary parts of slicer with hot soapy solution. *Be careful of the blade.*
8. Rinse with clean cloth dipped in clear water.
9. Sanitize stationary parts, wipe with sanitizer solution. Allow to air dry.
10. Reassemble.

PREVENT CONTAMINATION

UTENSILS
Wash and sanitize
after every use

CUTTING BOARDS
Wash and sanitize
after every task

HANDS
Wash frequently
and thoroughly

EQUIPMENT
Clean and sanitize
after every use

OBSERVE SANITARY RULES

EVITE CONTAMINACION

UTENSILIOS
Lávelos y desinféctelos
después de usarlos

TABLAS PARA PICAR
Lávelas y desinféctelas
cada vez que las use

MANOS
Láveselas con frecuencia
y cuidadosamente

EQUIPO
Lávelo y desinféctelo
despues de usarlo

OBSERVE NORMAS SANITARIAS

Using Thermometers*

Temperature is an integral part of food safety. Choosing a proper thermometer and using it correctly are important to prevent food safety hazards at critical control points.

Temperature control is important:

- To keep food out of the danger zone
- To document critical food temperatures in cooking, cooling, reheating, hot holding, and cold holding
- To monitor temperature control at receiving and delivery

Different types of thermometers have certain advantages or limitations as compared with other types. It is essential to know how to use each type of thermometer to record accurate and consistent temperature readings.

It is also necessary to clean and sanitize a thermometer after each use with a food item to prevent cross-contamination. To sanitize a thermometer, use an alcohol wipe or chlorine or quaternary ammonia at the recommended concentration for sanitizing equipment.

A thermometer should be calibrated weekly, and no less than once a month, or when it is dropped or exposed to extreme temperatures.

Training employees how to take temperatures and calibrate their thermometers is essential in the HACCP system. Employees will be responsible for temperature recording and taking corrective action if products are not at proper temperatures.

Types of Thermometers

Bimetallic Stemmed Thermometer

- This is the most common type of food thermometer.
- It measures temperature through a metal stem, just past the dimple (about 2 inches from the tip). This dimple must be in the middle of the product to record an accurate temperature reading of the product.
- The reading dial is at the top of the unit.
- There is a calibration nut just below the dial for adjusting the temperature.

---

*Adapted with permission from *Serving Safe Food Employee Guide, Second Edition.* Copyright © 1993 by the Education Foundation of the National Restaurant Association.

Digital Thermometer

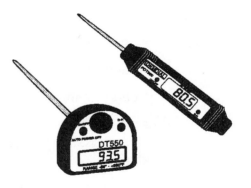

- This thermometer measures temperature through a metal tip at its end.
- It shows the temperature on a rapidly easy-to-read panel.
- Temperature calibration is set at the factory. When a digital thermometer is out of calibration, it must be standardized to calibrated thermometers or discarded.

Thermocouple

- This type of thermometer measures temperature electrically through a sensor in the tip.
- It shows the temperature on a digital readout panel.
- Temperature calibration can be adjusted manually.
- Many thermometers of this type are equipped with interchangeable temperature probes. These are especially practical to monitor the temperature of thin products, such as hamburgers and cold or hot spots on grills.

Time Temperature Indicators (TTI)

- These devices are usually found on packages and look like labels.
- They are used to monitor temperatures on sous vide, modified atmosphere packages, or cook-chilled foods.
- They have liquid crystals that change color or turn black if foods reach unsafe temperatures.

Equipment Thermometers

- Thermometers of this type are usually found in refrigerators, freezers, hot-holding equipment, and dishwashers.

Candy, Meat, and Deep-Fry Thermometers

- Such thermometers are used for only one type of food (candy, meat, or deep-fried foods).
- Never use mercury-filled or glass thermometers because they may break.

**How to Use a Thermometer
to Check Food Temperature**

1. Clean and sanitize the thermometer.
   Wash the thermometer in clean, soapy water.
   Sanitize with an alcohol wipe or in a sanitizer solution of chlorine or quaternary ammonia.

2. Insert the probe into the thickest part of the product.

3. Allow time for the thermometer to stablize.

4. Wash and sanitize the thermometer between each product checked.

5. Wash and sanitize before replacing the thermometer in its holder.

6. When checking a food delivery, it is recommended not to puncture through sealed packaging. Rather, place the thermometer between two products to check the temperature.

7. Calibrate each thermometer frequently (weekly or at least once a month, or after a thermometer is dropped). See the following information.

Calibrating a Probe Thermometer*

When calibrating a probe thermometer, handle it gently, because rough handling or dropping will cause it to lose calibration.

When: Calibrate each thermometer frequently.

How: 1. Fill a medium-sized glass with ice.
               Add water to ice.
               Place thermometer in glass of ice water.

            2. Wait three minutes.
               Stir water occasionally.

            3. After three minutes, thermometer should read 32°F.

Who: Each chef or cook doing food preparation of potentially hazardous foods should have access to a probe thermometer.

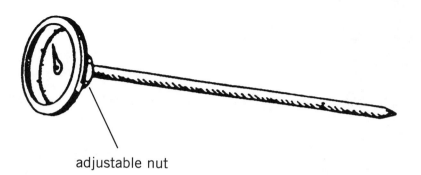

adjustable nut

*Denver Environmental Health Services, Consumer Protection/Food Safety.

Corrective Action

If a thermometer does not read 32°F after it has been in ice water for three minutes:

1. Leave it in the ice water.

2. Using pliers, 7/16 inch wrench, or an adjustable wrench, turn the adjustable nut on the back of the thermometer until the needle reads 32°F. It may be necessary to add more ice.

3. Wait three minutes, stir occasionally.

4. After three minutes, the thermometer should read 32°F. If not, repeat corrective action.

HACCP Project

In the following exercise, you will work through the second HACCP step.

1. Identify potentially hazardous foods (PHF).

2. **Identify critical control points.**

3. Establish control procedures to guarantee safe food.

4. Establish monitoring procedures.

5. Establish corrective action.

Review the fried chicken breast recipe, then go to Step 2 on page 89.

FRIED CHICKEN BREAST

Items:

Boneless Chicken Breast
Flour
Salt
White Pepper
Eggs
Milk

Preparation:

1. *Mix* flour, salt, and pepper.
2. *Mix* egg with milk; blend well.
3. *Dip* chicken in egg wash, then in flour.
4. *Deep fat fry* at 375°F for approximately 4 minutes.
5. Remove, place on wire rack in sheet pan.
6. *Bake* in 375°F oven for approximately 20 minutes until done.
7. Remove and *chill*.
8. *Reheat* to serve.

HACCP Step 2: Identify Critical Control Points

FRIED CHICKEN BREAST

A. Review the Fried Chicken Breast recipe.

B. Make a simple flow chart of the recipe on the previous page, charting the 9 preparation steps from SOURCE through the preparation process to SERVING. See examples in this section.
    Hint: Most of the key words are in the *Preparation* part of the recipe.

C. Mark the CCP points in the Fried Chicken Breast flow chart.

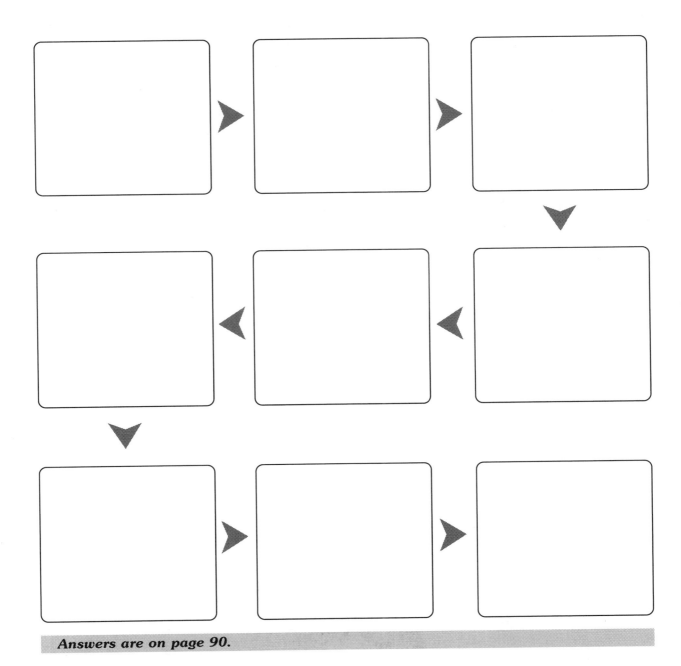

Answers are on page 90.

ANSWERS:

HACCP Step 2: Identify Critical Control Points

A. Review the Fried Chicken Breast recipe.

B. Make a simple flow chart of the recipe on the previous page, charting the 9 preparation steps from SOURCE through the preparation process to SERVING. See examples in this section.
    Hint: Most of the key words are in the *Preparation* part of the recipe.

C. Mark the CCP points in the Fried Chicken Breast flow chart.

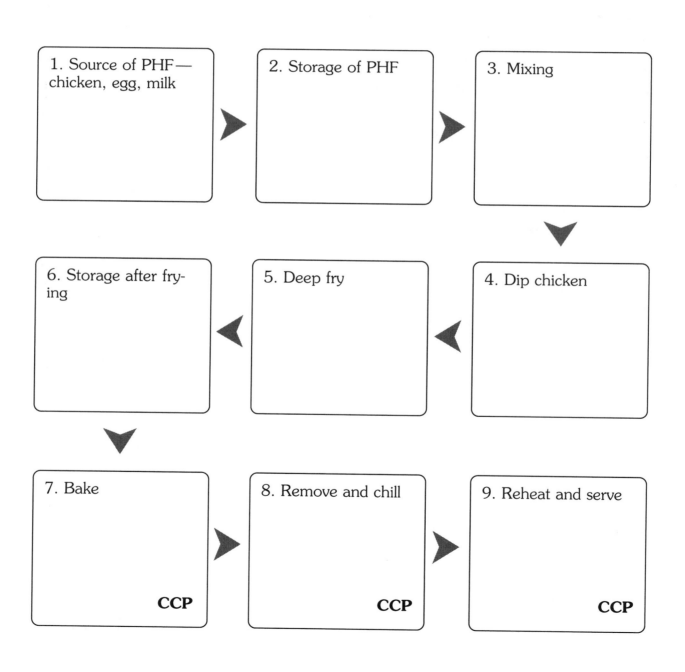

1. Source of PHF—chicken, egg, milk	2. Storage of PHF	3. Mixing
6. Storage after frying	5. Deep fry	4. Dip chicken
7. Bake CCP	8. Remove and chill CCP	9. Reheat and serve CCP

STEP 3

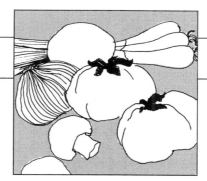

Establish Control Procedures

Critical limits are defined as the criteria that must be met for each preventive measure associated with a critical control point (CCP). Critical limits may be set for preventive measures such as:

Time

Temperature

pH

Water

> *I*f we did the things we are capable of doing, we would literally astound ourselves.
>
> *–Thomas A. Edison*

APPLICATION OF STEP 3

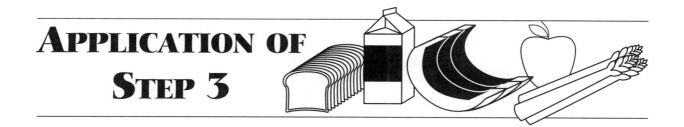

Establish Control Procedures

Enforce employee hand washing, hygiene, and serving practices.

Enforce proper cleaning and use of sanitizer solutions.

Write recipes with control points.

    Storage
    Thawing
    Prepping
    Cooling
    Reheating
    Holding
    Serving

Critical Limits

The critical limits over which we have the most control are time, temperature, pH, and water activity.

Time

Time and temperature work together to become the time bombs of foodborne illness. Under the most favorable conditions, bacteria double every 20 minutes. Controlling time in the danger zone (see page 105) is a very effective control limit.

Temperature

Temperature is one of the factors that can be used to "manage" the number of organisms that may be present in a food product. Temperature is a well-known management tool to keep bacteria levels low.

pH

The pH of a food product is a measure of its acidity or alkalinity.

A solution with a pH of 7.0 is considered neutral. As the product pH moves below or above the pH that is ideal for the growth of a particular organism, the bacteria take longer to adjust to their surroundings and the microorganisms grow more slowly.

Many foods are naturally acidic; that is, the pH is below 7.0. Some foods are quite acidic and have a pH below 4.6. Examples are vinegar, mayonnaise, fruits, pickles, and yogurt.

At or below 4.6 PH, disease-causing organisms do not grow or grow very slowly. This does not kill the organisms but only retards growth. Spoilage organisms, however, may grow at these low pH values and can slowly change a food's taste and appearance.

Water Activity

Microorganisms need water to grow. Because bacteria cannot take their food in a solid form, they must receive their nutrients in some kind of water solution. This solution is described as "water activity," which means the amount of water available for growth. Solutes (salts and sugars), as well as drying, decrease the available water and can reduce bacterial growth.

Critical Limits*

Each standard should be something that can be immediately monitored — by measurement or observation. Standards (critical limits) for CCPs must be as specific as possible.

- **Temperature**
- **Time**
- **pH**

Preventive Measures

Preventive measures should be detailed in standard operating procedures.

- **Cleaning and sanitizing**
- **Washing hands**
- **Covering food containers**
- **Separating raw and cooked products**

*Adapted with permission from *Managing a Food Safety System* seminar. Copyright © 1992 by the Education Foundation of the National Restaurant Association.

Control Procedures*

Once the critical control points (CCPs) are determined, critical limits or standards can be established to reduce or eliminate potential hazards.

Sometimes you can reduce or eliminate hazards by setting specific time or temperature limits. For example, rapid growth of harmful bacteria is a hazard in many perishable foods. This is especially true for meat, poultry, and seafood. For foods containing meat, poultry, and seafood, receiving, chilling, and chilled storage are CCPs—stages in the food preparation process in which bacterial growth can be controlled. To control bacterial growth during these stages, temperature limits can be set. Meat, poultry, and seafood must be at or below 40°F to control harmful bacterial growth.

In other steps of the food preparation process critical limits are set to "kill" the bacteria. This is why you will notice that different final cooking temperatures are mandatory for meat, poultry, and fish products. For example:

Chicken	165°F	kills salmonella
Hamburgers	155°F	kills *E. coli* 0157:H57
Pork	155°F	kills trichinosis

Sometimes you can reduce or eliminate hazards by setting standards for products received. For most products, receiving will be a control point. However, certain products such as seafood or modified atmosphere products (MAPs) can be hazardous because further processing steps will not eliminate the risk, and receiving will become a critical control point. Controls may include rejecting foods that show evidence of time/temperature abuse or that are not received from approved sources. Fish and seafood from unapproved sources carry a higher risk of contamination and seafood toxins. MAPs, if time/temperature abused or out of date, may promote the growth of pathogens.

When controlling critical items, keep in mind that bacteria need four things in order to grow:

- The right amount of time
- The right temperature
- The right pH (acidity or alkalinity)
- The right amount of moisture

Keeping these factors within certain limits will prevent bacterial growth.

---

*Ensuring Food Safety—The HACCP Way: An Introduction and Resource Guide for Retail Deli Managers, University of California Cooperative Extension, Davis, California.

The following are examples of critical limits or standards to reduce or eliminate potential hazards at CPs and CCPs (proposed new code temperatures).

*CPs or CCPs: Receiving**

- Perishable products at or below 40°F.
- Frozen foods at 0°F or below.
- Purchase only from approved sources.

CCPs: Cooking, Reheating, and Hot Holding

- Cook poultry to at least 165°F (15 sec.).
- Cook pork to at least 155°F (15 sec.).
- Cook roast beef to at least 145°F (15 sec.).
- Cook hamburger to at least 155°F (15 sec.).
- Cook fish to 145°F (15 sec.).
- Reheat all foods rapidly to at least 165°F (15 sec.) within two hours.
- Hold all hot foods at 140°F or higher.
- Prepared foods held at room temperature for two hours should be destroyed.

CCPs: Chilling and Chill Storage

- Cool all foods from 140°F to 70°F in two hours, and from 70°F to 40°F within four hours.

*CPs: Food Handling (preventive measures)***

- Thoroughly wash vegetables in clean, cold water.
- Use proper hand washing techniques.
- Cover and protect open cuts and scratches.
- Stay home when sick.
- Handle cooked foods only with clean gloves or utensils.
- Follow proper dish washing and sanitizing techniques.
- Use clean and sanitized equipment and utensils.
- Work with small amounts of product to maintain temperatures.
- Maintain sanitizer buckets with fresh solutions. Check concentration with test strips.

---

* Based on product. If further processing steps can control, eliminate, or reduce the hazard, receiving is a control point. If further processing will not prevent a hazard, such as with seafood or MAPs, receiving is a critical control point.

** Food handling practices are control points because they are difficult to measure, monitor, and document. The exception may be in processing plants where the process becomes part of the established procedure and can be documented.

Temperatures for Potentially Hazardous Foods

This simplified flow chart for turkey illustrates some standards identified at control points in the food flow. Observe that receiving, cooking, and cooling identify temperature as a control point. In the cooling process, time is also identified as a control point.

Inspection, sanitizing, and hand washing are preventive measures, as they are difficult to measure, monitor, and document all of the time.

Simplified HACCP

PHF: **Turkey**
(Potentially Hazardous Food)

CONTROL POINT	CRITICAL LIMIT (Standards)
Source	USDA inspected product.
Receiving	No evidence of thawing, received at 0°F or below.
Cooking	Center temperature 165°F.
Boning	Clean, sanitized utensils. Workers wash hands before handling cooked product.
Cooling	Total exposure time in danger zone not to exceed six hours. Center temperatures from 140°F to 70°F within two hours and from 70°F to 40°F within four hours.

Proposed New Food Code Temperatures

The proposed new food code temperatures have been used in this book. This proposed code is based on the latest scientific evidence and takes a more progressive approach to food safety. Because it may be some time before the new code is adopted, it is best to check the regulations in your area.

The lower limit of the danger zone is changed in the proposed code from 45°F to 41°F. By controlling *Listeria,* at refrigeration temperatures all other organisms are controlled for cold holding. (See the Fact Sheet on *Listeria,* page 238, in the Foodborne Illness section of this book). The temperature of 41°F is based on the point where 5°C falls on the Fahrenheit scale. The more progressive companies, such as General Mills, have been using 40°F instead of 41°F in their standards. It is believed most of industry will follow that lead for simplicity.

The proposed code raises the final cooking temperature of some foods and adds a 15-second holding standard to all final cooking temperatures. The pasteurization standard for foods is based on the optimal growth potential of microorganisms found in particular food products. (See the Fact Sheet on *Salmonella,* Pasteurization Standards, on page 240 in the Foodborne Illness section of this book). If salmonella is controlled through the cooking and reheating process, all other vegetative types of bacteria are controlled.

The proposed cooling code may be controversial. The existing code for product cooling is from 140°F to 45°F in four hours or less. The proposed cooling code allows a two-tiered cooling standard; cooling from 140°F to 70°F in two hours, and from 70°F to 41°F in four hours. This two-tiered cooling standard is based on the growth curve of *C. perfringens*. If *C. perfringens* is controlled through the cooling process, all other microorganisms are controlled. (See the Fact Sheet on *C. perfringens*, Optimal Growth Potential, on page 239 in the Foodborne Illness section of this book).

Modification of Posters to Meet Codes

Many of the posters and training aids in this book are ready for use as presented. However, you may wish to modify pages to meet codes in your area. Certain sections of *The HACCP Food Safety Manual* can be made available on computer disk or on hard copy. For further information and prices, contact the author at Hospitality Personnel Services, 190 East Ninth Avenue, Suite 190, Denver, CO 80203. The telephone number is (303) 830-6868.

PROPOSED NEW FOOD CODE		
	Current Standard	**Proposed New Code**
Danger Zone	45°F–140°F	41°F–140°F
Cold Holding	45°F or Below	41°F or Below
Hot Holding	140°F or Above	140°F or Above
Cooling	140°F to 45°F/4 hrs.	140°F to 70°F/2 hrs. and 70°F to 41°F/4 hrs.
Reheating	165°F	165°F (15 sec.)
Cooking:		
Pork	150°F	155°F (15 sec.)
Poultry and Stuffed Meats	165°F	165°F (15 sec.)
Beef, Fish, and Eggs	140°F	145°F (15 sec.)
Ground Beef and Pork		155°F (15 sec.)
Microwave Cooking		+25°F to final cooking temp.
Labeling	Date	Date and Time of Prep.

LIBRARY LRS

Control Temperatures

1. Cold Food Holding at 40°F or below

2. Hot Food Holding at 140°F or higher

3. Roast Beef 145°F (15 sec.)
   Beef roast (rare) at 140°F (12 min.)
   Beef roast (rare) at 130°F (121 min.)

4. Pork Products 155°F

5. All Leftovers 165°F

6. Poultry and Stuffing Products 165°F

7. Fish and Shellfish 145°F

8. Fresh Egg Products 145°F (3½ min.)

9. Hamburger Patties 155°F (until no longer pink and juices are clear)

FOOD FLOW CHART
Beef Stew

RECEIVING AND STORAGE

Receive and store beef at 40°F or below.
Store on lower shelf.

PREPARATION

Use cleaned and sanitized utensils.
Pull and cube one roast at a time.
Refrigerate each batch as completed.
Wash hands before handling vegetables.
Use clean and sanitized utensils for vegetables.

COOKING

HOLDING

COOLING

REHEATING

Review the Beef Stew Flow Chart; mark the CCPs. A CCP is a step that, if not performed correctly, may result in a hazard that would not be prevented by a following step. Write the critical limit at each CCP.

Answers are on page 102.

Beef Stew

RECEIVING AND STORAGE

Receive and store beef at 40°F or below.
Store on lower shelf.

PREPARATION

Use cleaned and sanitized utensils.
Pull and cube one roast at a time.
Refrigerate each batch as completed.
Wash hands before handling vegetables.
Use clean and sanitized utensils for vegetables.

COOKING

CCP Cook to minimum of 165°F.

HOLDING

CCP Hold at 140°F or above.

COOLING

CCP Cool rapidly from 140°F to 70°F in two hours or less and from 70°F to 40°F within four hours or less.

REHEATING

CCP Reheat rapidly to 165°F (15 sec.) within two hours.

FOOD FLOW CHART
Cold Seafood Salad

RECEIVING AND STORAGE

Frozen pre-cooked crab received and stored at _____°F or lower.
Frozen raw shrimp received and stored at _____°F or lower.
Purchase from approved sources (Certificate of Conformance).
Whole-shell eggs received at _____°F or lower.

THAWING

Thaw crab meat under cold running water (_____°F or below) in two hours or less.

COOKING

Cook whole-shell eggs in boiling water for 10 minutes until yolks reach internal temperature of _____°F (15 sec.).
Cook frozen shrimp until it reaches an internal temperature of _____°F (15 sec.).

COOLING

Cool eggs rapidly from _____°F to _____°F in ice water.
Cool shrimp rapidly from _____°F to _____°F in ice water.

PREPARATION

Wash all salad ingredients.
Assemble salads following recipe food safety instructions.

HOLDING AND DISPLAY

Wrap plated salads with plastic film.
Hold salads at _____°F or lower.
Store separately from and above raw _____.
Hold for four hours from beginning of preparation and then destroy.

Review the Seafood Salad Flow Chart. Mark the CCPs and write the critical limits. Review the recipe again. Could the number of CCPs selected for monitoring be reduced with a few procedural changes?

Answers are on page 104.

FOOD FLOW CHART
Cold Seafood Salad

RECEIVING AND STORAGE

CCP

Frozen pre-cooked crab received and stored at 0°F or lower.

Frozen raw shrimp received and stored at 0°F or lower.

Purchase from approved sources (Certificate of Conformance).

Whole-shell eggs received at 45°F or lower.

THAWING

Thaw crab meat under cold running water (70°F or below) in two hours or less.

COOKING

CCP

Cook whole-shell eggs in boiling water for 10 minutes until yolks reach internal temperature of 145°F (3½ min.).

Cook frozen shrimp to an internal temperature of 145°F (15 sec.).

COOLING

Cool eggs rapidly from 140°F to 40°F in ice water.

Cool shrimp rapidly from 140°F to 40°F in ice water.

PREPARATION

Wash all salad ingredients.

Assemble salads following recipe food safety instructions.

HOLDING AND DISPLAY

CCP

Wrap plated salads with plastic film.

Hold salads at 40°F or lower.

Store separately from and above raw PHF.

CCP

Hold for four hours from the beginning of preparation, then destroy.

LIMIT TIME THAT FOOD IS IN THE DANGER ZONE

Between 40°F and 140°F (or between 5°C and 60°C)

THAWING

PREP TIME

COOLING

REHEATING

NO MORE THAN A COMBINED TOTAL OF FOUR HOURS FOR ALL PROCEDURES

Temperatures for Potentially Hazardous Foods (PHF)

165° • Reheat all leftover foods to 165°F (15 sec.). Cook all poultry to 165°F (15 sec.) (internal product temperature).

155° • Cook pork, pork products, and ground beef to 155°F (15 sec.).

145° • Internal temperatures for roast beef, fish, and eggs (15 sec.).

140° • Hold all hot PHF at 140° F or above. **Cool quickly** (from 140°F to 70°F in two hours and from 70°F to 40°F in four hours).

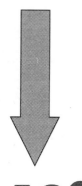

1. Shallow containers 2 inches deep in refrigerator or freezer.
2. Quick chilling in ice and stirring.
3. Check with probe thermometer. Record on food temperature log.

40° Hold all cold PHF at 40° F or below.

Potentially hazardous foods (PHFs) include cooked rice, beans, baked or boiled potatoes, meat, poultry, fish, shellfish, milk, milk products, and eggs.

Note: Best refrigerator temperature is 37°F.

For microwave cooking, add 25°F to the final cooking temperature.

Cook Ground Beef to 155°F*

MINIMUM HAMBURGER COOKING TEMPERATURE

Freezing hamburger patties does NOT kill bacteria. Food poisoning outbreaks in 1993 in northwest United States from contaminated, undercooked hamburgers caused:

604 cases of food poisoning
43 people to be hospitalized
4 children's deaths

Denver Department of Public Health/Consumer Protection.

Rapid Cooling Means Safe Food

The temperature danger zone for food products is 40°F to 140°F. Many bacteria multiply rapidly in this climate. Improper cooling is the number one cause of foodborne illness. (See "To Cool" on page 111.)

The quicker we can cool our products, the less time we allow the conditions for bacterial growth in the danger zone. Remember, we must throw out food if it is in this temperature range longer than four hours. (Refer to "Bacterial Growth" on page 243 in the section "Foodborne Illness.")

For years we assumed the refrigerator would be sufficient to cool down food. We now know from repeated tests that most refrigerators are severely lacking in capacity to rapidly cool food safely. In fact, very hot food and large quantities of warm food will actually raise the temperature inside a refrigerator. This may pose food safety problems for other food products in the refrigerator.

Get your staff involved. Conduct cool-down experiments. Start by recording cool-down temperatures of several products in the refrigerator as you normally would. Record time and temperature until a product reaches 40°F. Then have your staff experiment with different methods of cooling; again, record time and temperature. They can determine which methods work best for different products.

There are several factors that can help speed the cooling of foods, which are discussed in the following paragraphs.

Containers

Aluminum metal chills fastest, followed by stainless steel. Glass and plastic are poor containers to use for cool down, as they transfer heat out of the food product very slowly. The shallower the pan, the faster the food cools.

Volume

Small batches cool faster. Small batches in shallow metal pans chill even more rapidly. Reduce larger quantities of heated foods to smaller quantities. (Refer to "Possible Cooling Solutions" on page 110.)

Stirring

Stirring increases cooling. Speed up the process by occasionally stirring a product while chilling. Stir the food product every 15 minutes or each time you walk by it.

Air Circulation

Uncovered foods chill faster. No matter how the container is covered—foil, film, another pan, or a lid—cool air cannot effectively reach the food when it is covered. Rapidly chill, then cover.

When cooling in the refrigerator, the product depth should be no greater than 3 inches. Dense products such as beans or rice should cool in shallow pans with a product depth of 2 inches or less.

Stacked pans block air circulation. When stacked, the rolling rack or pile of pans becomes one big hot unit. Air can reach only the outside of the stack.

Maximize the air flow in the refrigerator. To maximize air flow, use commercial wire racks and do not block the blower. Since the blowers are located near the ceiling, food cools first on the top shelves and farthest from the door. To maintain even temperatures in a walk-in refrigerator, do not leave the door open unnecessarily, or use plastic draft strips.

Ice

To quick-chill, use ice or ice baths. Ice is the most convenient chiller in a commercial kitchen. Ice can be substituted for water in a recipe and added at the end of the cooking process (soups, stews, or gelatins) to cool the product rapidly. For the most effective use of ice, cool the product by stirring to 140°F (the temperature at which food begins to enter the danger zone), then add ice to chill rapidly to 40°F.

The ice-bath method drops temperatures of hot products most effectively. Use a sink or container large enough to hold the pan of hot food and three times the amount of ice as product to be chilled. Nestle the food pan in the ice, covering the bottom and sides of the pan. When the volume of ice is greater than the volume of water, it is 70 percent more effective in cooling the product. Remember these secrets to rapid cooling (see "To Cool" on page 111).

Use an aluminum or stainless steel shallow pan.
Keep the quantity small.
Uncover and stir occasionally.
Replace the ice when it melts to half of the original volume.
Refrigerate food when chilled.
Cover when product is cooled to 40°F.

Other Factors

As an operation grows, so may your needs for more refrigeration or for the new blast chillers. If your foodservice is adaptable to pre-cooked, pre-portioned foods, you might consider adapting your menu to some of these items to eliminate the risks of food handling time and contamination.

POSSIBLE COOLING SOLUTIONS

1. USE MORE "COOK AND SERVE." (Prepare and immediately serve.)

   –or–

2. REDUCE LARGE QUANTITIES OF HEATED FOODS TO SMALLER PORTIONS. Divide large pot of food into several smaller pots or shallow pans; cut large roasts into smaller pieces to cool more quickly.

3. USE ICE BATHS TO COOL. Completely surround container with an ice-water mixture and stir every 15 minutes.

4. WITHHOLD WATER DURING COOKING from stews and soups. THEN ADD ICE AT END TO REPLACE THE WATER.

5. USE SHALLOW METAL PANS (2 inches or less). Store on the upper shelf of cooler. Leave uncovered, stirring every hour if possible until product reaches 40°F.

6. DIVIDE FOOD INTO PLASTIC BAGS; seal, surround bags with ice to cool.

7. PURCHASE A BLAST CHILLER.

8. ADD MORE WALK-IN REFRIGERATION SPACE.

To Cool
Potentially Hazardous Foods

from

within six hours (from 140°F to 70°F in two hours and from 70°F to 40°F in four hours):

1. Place *uncovered* in a refrigerator or freezer (on top shelf) in a 2-inch shallow metal pan and stir every 15 minutes.

or

2. Quick-chill in an ice water bath. Stir and cool to 40°F before putting in refrigerator.

Rapid Cooling Considerations

Cooling Containers for Foods

Aluminum	excellent
Stainless Steel	good
Plastic or glass	poor

COOLING IS MORE IMPORTANT THAN COVERING.

WHEN THE ICE VOLUME IS GREATER THAN WATER, IT IS 70% MORE EFFECTIVE.

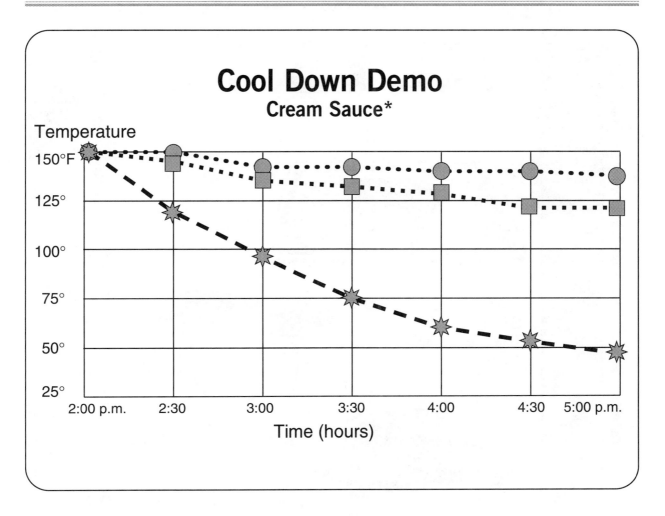

Cool Down Demo
Cream Sauce*

*Cream sauce cooled in a two-gallon stainless steel round container

⬤ Cooled in refrigerator, no stirring or ice

⬛ Cooled in ice bath without stirring

✹ Cooled in ice bath stirred every 15 minutes

Conduct your own cool-down demonstration.
For faster results use 2-quart stainless steel and plastic containers. Instant rice or water works well for class demonstrations. If a refrigerator isn't available, an alternative cool down test would be to cover one container with plastic film and use one plastic container instead of metal.

Add 2 cups of water, all at the same temperature, 140°F or hotter.

Repeat the cool-down examples above.

Record your temperatures.

Graph your temperatures.

KEYS TO

COOL FOODS QUICKLY
Use an ice bath

STORE LEFTOVERS SAFELY
Use small, shallow containers

CUT MEATS DOWN TO SIZE
Slice to 3 inch or less in-thickness

165° F

REHEAT FOODS TO SAFE TEMPERATURES
Reheat to 165°F

SAFE COOLING AND REHEATING

CLAVES PARA

ENFRIE LOS ALIMENTOS RAPIDAMENTE

Use un bāno de hielo

GUARDE LO QUE SOBRE EN RECIPIENTES SEGUROS

Use recipientes pequeños y llanos

CORTE LA CARNE EN PEDAZOS

Córtela en tajadas o trozos de 3 pulgadas o menos

RECALIENTE LOS ALIMENTOS HASTA QUE LLEGUEN A LA TEMPERATURA CORRECTA

Recaliéntelos hasta que lleguen a 165°F

ENFRIAR Y RECALENTAR LOS ALIMENTOS

Safe Ways of Thawing Frozen Foods

Thawing methods must be planned to minimize the time food is in the danger zone and to prevent cross-contamination. Four acceptable methods of thawing food are:

1. Thaw under refrigeration temperatures of 40°F or lower. This method requires advance planning of usually of 24 to 48 hours, depending on the size of the item.

Thawing foods require adequate refrigeration space. Place raw potentially hazardous foods on the bottom shelf. All thawing foods should be placed in pans to prevent dripping and spilling onto other foods.

2. Thaw under potable running water at 70°F or below for no more than two hours. This method is not recommended for large food items that will not thaw in two hours.

Cool running water is required so as to prevent rapid bacterial growth and to allow loose food particles to overflow.

Use a clean, sanitized prep sink or pot.

Prevent cross-contamination by not mixing raw food items in the same sink or container. Different products have different pathogens and require different "kill" temperatures, such as chicken at 165°F and fish at 145°F.

Wash, rinse, and sanitize all equipment, sinks, and utensils after food is thawed.

3. Thaw by cooking frozen product to 140°F or above and all leftovers to 165°F or above.
4. Thaw in a microwave oven. This method is recommended for small food items or single-service items. For food thawed in a microwave:

   - Transfer immediately to conventional cooking to complete the cooking process (or)
   - Continue with uninterrupted cooking in the microwave.

The new code for microwave cooking will add 25 degrees to a product's final cooking temperature. Rotate and stir during cooking, cover to retain moisture, and allow a two-minute standing period for even temperature.

Critical Control Point—Thawing

Four Safe Ways of Thawing Frozen Foods*

Refrigerator (40°F)

Cold Running Water
at 70°F or below in two hours
or less

Cooking Process
(Continue with no interruption
the thawing process to completion
of the cooking process.)

Microwave Oven
(Add 25°F to product's
final cooking temperature.)

Question: Why not thaw frozen foods at room temperature?

Answer: Thawed portions of potentially hazardous foods will support bacterial growth. Prolonged thawing of these foods at room temperature can result in excessive bacterial growth and an unsafe product.

*Adapted with permission from *Applied Foodservice Sanitation Visual Aid Program.* Copyright © 1991 by the Education Foundation of the National Restaurant Association.

Purchasing, Receiving, and Storage of Food Products

Purchasing

Most foodservice operators rely on the food purchased to be inspected before it comes into the operation. Purchasing from approved sources is important in establishing your HACCP system. Some foods, such as seafood, relate to biological and chemical concerns. Seafood from contaminated waters and some fish species can cause foodborne illness. Working closely with a highly reputable suppliers and their sources can help prevent problems.

All food products must be purchased from approved sources. Work with your suppliers so you can receive deliveries when employees have time to inspect and properly store the food products delivered.

Receiving

For some foods, receiving can be a critical control point if further processing would not be a "kill" or control step to prevent a foodborne illness. Such foods may include raw seafood, and modified atmosphere packaged foods (MAPs).

Set up procedures for inspection, standards for acceptance, and procedures for rejection if products do not meet standards. Train employees in inspection procedures.

Storage

Proper storage can prevent bacterial growth and cross-contamination. When food is properly stored it keeps longer, reduces waste, and can prevent dangerous foodborne illness. Rules for safe storage include the following:

- Date foods when received. Use the oldest products first.
- Keep potentially hazardous foods (PHFs) out of the danger zone (40°F–140°F). Immediately store PHFs in the refrigerator or freezer upon receiving or during the preparation stage.
- Prevent cross-contamination during storage. Store raw foods below cooked foods. Store raw meats on the lowest shelf.
- Transfer opened packaged foods to clean, sanitized food storage containers with tight lids. Label and date foods removed from their original packaging.
- Maintain clean, dry storage areas and clean storage and transporting equipment.

Applying HACCP to Storage

If you were storing these food items on a shelf in your refrigerator, how would you store them from top to bottom?

Raw Beef

Raw Pork

Raw Chicken

Fresh Vegetables

Cooked Casserole

Frozen Fish

Hamburger

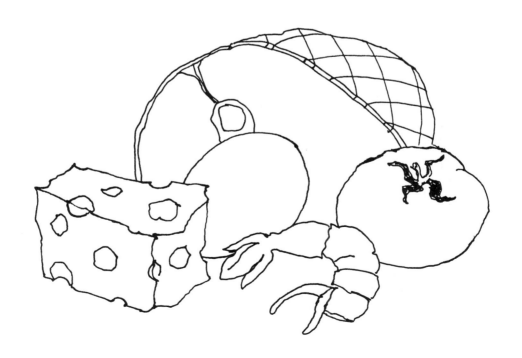

Answer to HACCP Storage

You will notice that the raw products' final cooking temperature determines the storage level if products could be stored only on a series of shelves.

Cooked Casserole
Fresh Vegetables
Frozen Fish
Raw Beef
Raw Pork
Hamburger
Raw Chicken

Proper Handling of Prepared Foods

Holding and Displaying

Before prepared foods are held for service or displayed, establish temperature controls. Certain foods require minimum cooking temperatures to kill pathogens. Otherwise, foods held at favorable conditions for long periods may promote pathogenic growth to levels that can cause foodborne illness.

Often a foodservice worker is tempted to maintain food quality by keeping temperatures low enough so that food doesn't dry out rapidly, or by keeping a roast beef rare by holding temperatures low. When conflicts occur between culinary and food safety concerns, food safety must be the deciding factor.

A foodservice manager needs to establish standards and practices so as to reduce the conflict between culinary and food safety issues. The following practices can help to resolve both concerns:

- Prepare food as close to service as possible.
- Put mixtures in smaller pans and stagger cooking through the meal period. Refrigerate uncooked/unheated pans and/or ingredients.
- Control the batch cooking on an as-needed basis. Establish holding standards to preserve quality.
- Maintain hot food temperatures at 140°F or higher.
- Maintain cold food temperatures at 40°F or lower.
- Verify food temperatures with a calibrated thermometer.
- Stir soups, gravies, and sauces to maintain even temperatures.
- Add hot liquids to maintain consistency and prevent drying out.
- Cover food to retain heat and prevent drying out.
- Never use hot-holding equipment or steam tables to heat food. They are not designed for rapid reheating of food.

Serving

Serving is the last crucial step before food is received by the customer. When food is ready to be served, there are no "kill" steps to destroy bacteria, only control steps. Food handlers must be made aware of how they can control contamination of cold and ready-to-serve foods:

- Hand washing must be monitored and enforced, before the food handler is allowed to prepare or serve food.
- Food handlers must be made aware of poor personal habits, such as touching the mouth, face, or hair; wiping hands on a dirty apron or cloth that may be a source of contamination.

Employees must be trained:

- Not to touch plates, utensils, drinking glasses, or cups where the customer's food or mouth will come in contact with the surface.
- When dishing a customer plate, to wipe drips with a clean cloth or fresh paper towel. The counter cloth should not be used, as it may contain contaminants.
- To use sanitized utensils to handle food, not bare hands.
- To use proper serving utensils with long handles so as to prevent handles coming in contact with food.
- When scoops and ladles are not in use, they should be stored in the food with the handle out. Tongs must be stored on a dry, clean surface or in a separate pan.
- Some operations require the use of gloves for handling of food that will not be cooked and for which the use of utensils is not practical, such as sandwich making. Gloves must be changed under the same conditions as hand washing.
- Employees must be aware of related duties that require hand washing before continuing with service, such as:
    Picking up an item from the floor
    Handling soiled dishes and linens
    Answering the telephone; handling cash

Management should schedule duties and staffing to minimize tasks that can be sources of contamination, such as bussers handling soiled dishes and servers handling only food and clean dishes.

- Self-service stations have increased risk of contamination by the customers and require attention to maintain proper temperatures.
    Use sneeze guards to protect food.
    Provide sufficient long-handled utensils.
    Don't overfill containers so that food comes in contact with handles.
    Require customers to use a fresh plate with each return.
    Discourage eating or picking with hands at the station.

Leftovers/Reheating

Once the meal period is over, the foodservice manager must make certain decisions about leftovers. Any leftovers must be held and reheated with special care to prevent bacterial growth. Leftovers have several food safety implications such as:

- Once reheated, leftovers will have passed through the temperature danger zone three times.
- Leftovers have a high risk of contamination; therefore they should be kept to a minimum and handled with special care.
- Highly susceptible foods such as custards, puddings, and creamed casseroles must not be reheated. If these items are leftovers, they must be thrown out.
- All potentially hazardous leftovers should be chilled to 40°F within two to six hours. Reduce time for chilling if the food has already been in the danger zone.
- Proper cooling procedures must be followed. Choose a method appropriate for the type of product.

    A very dense product should be divided into shallow pans with product thickness less than 2 inches.

    Cut large pieces of meat into smaller pieces; place in shallow pans separated for good air circulation or surround with bags of ice.

    Use ice baths and stir every 15 minutes.

    Divide into smaller metal containers and use ice baths.

    Divide into plastic bags, seal, and surround with ice.

- Store leftovers uncovered while cooling to 40°F; once cooled, cover and place on upper shelves of refrigerator.
- Use sanitized calibrated thermometers to verify temperature.
- Label prepared foods with date and time.
- Reheat only that food that has been properly cooled and refrigerated for two days or less.
- Rapidly reheat foods to 165°F or higher within two hours.
- Never mix leftovers with fresh product.
- Once leftovers are reheated and held for service, they should be thrown out.

Eggs and Egg-Based Products

Outbreaks of salmonellosis have been traced to clean, whole, uncracked-shell eggs contaminated with *Salmonella enteritidis*. Whole-shell eggs are now classified as potentially hazardous foods by the FDA.

The following guidelines, which treat eggs as potentially hazardous foods during storage, handling, preparation and service, must be followed to prevent the possibility of foodborne illness:

- Store eggs at refrigerated temperatures of 40°F or lower.
- When using eggs for menu service, it is best to keep them refrigerated and pull as needed. If this is not feasible, pull one layer at a time if used quickly, or ice eggs in a pan.
- Cooked eggs must be held at 140°F or higher.

Proper Handling of Prepared Foods

- Label foods with their preparation date and time.

- Rapidly cool in 2-inch shallow metal pans to 40°F or lower within six hours (140°F to 70°F within two hours and from 70°F to 40°F within four hours).

- When Cooling:
    Never stack shallow pans one on top of the other.
    Allow air to circulate around foods to be chilled.

- Separate hot foods from cold.

- Always cover leftovers after they are cooled.

- Never mix leftovers with fresh product.

- Rapidly reheat to 165°F within two hours.

- Never use hot food holding equipment (chafers and steam tables) to reheat foods or cold-holding equipment to cool down foods (thermal holding cabinets).

- Use sanitized calibrated thermometers to monitor temperatures.

- Never reuse foods that have been previously served to customers, including breads, butter, sauces, dressings, and chips.

- Cook eggs thoroughly until both the yolk and white are firm, not runny.
    Scrambled—1 minute on cooking surface of 250°F
    Poached—5 minutes in boiling water
    Soft-boiled—7 minutes in boiling water
    Sunnyside—7 minutes on cooking surface of 250°F, or covered 4 minutes at 250°F
    Fried, over easy—3 minutes at 250°F, then turn and fry another 2 minutes on the other side.

- Avoid pooling raw eggs for holding. Eggs may be pooled in small quantities for immediate cooking and serving.

- For lightly cooked egg items, such as custards, French toast, mousse, and meringues, pasteurized eggs should be used.

- Avoid raw egg menu items. Review menus, recipes, and preparation procedures using raw eggs. Pasteurized eggs may be substituted for Caesar salad, hollandaise and bearnaise sauces, eggnog, ice cream, and egg-fortified beverages.

- Pasteurized eggs require the same time and temperature handling as other potentially hazardous foods.

- Wash hands with hot, soapy water before and after handling eggs and egg products.

- Wash and sanitize utensils, equipment, and the work area after handling eggs and egg products.

- Do not reuse a container that has held a raw-egg mixture. Use a clean, sanitized container for each batch.

123

Applying HACCP Control Procedures to a Recipe

In the first three HACCP steps, hazards have been identified, controls determined, and control procedures established. Now follow the food flow through the operation. Look for problems and practices that have been identified in the first three HACCP steps as food passes through each phase in the flow.

The following lists identify control points at each phase in the food flow. Some of the steps are control points and others are critical control points (CCPs). At the CCPs, determine critical limits for each product and the procedures that are to be monitored and verified. At the control points (CPs), establish standard operating procedures (SOPs) and handling practices to control any potential hazards. (See the example of the Seafood Salad Flow Chart on pages 153–158.)

Use the following lists as guidelines of what observations should be made and what to monitor as a food is flow charted.

The examples in the following paragraphs show how you can change your current recipe system into the HACCP system. Writing control procedures into the recipe also teaches your staff about food safety.

*Food Flow Control Points**

RECEIVING

- Time–Temperature
    - Refrigerated foods at 40°F or below
    - Frozen foods at 0°F or below
- Organalephic Evaluation
    - Odor, sight, touch
- Documentation
    - Shellfish tags, saved for 90 days
    - Certificate of Conformance (contract with supplier for no species substitution and safe source)

STORAGE

- Time–Temperature
    - Refrigerated foods at 40°F or below
    - Frozen foods at 0°F or below
    - Manufacturers' storage guidelines
- FIFO (First In–First Out)
    - Code Dating
- Prevention of Cross-Contamination
    - Storage procedures
    - Clean, wrapped containers
    - Containers for transporting food

---

*Adapted with permission from *Managing a Food Safety System* seminar. Copyright © 1992 by the Education Foundation of the National Restaurant Association.

THAWING

- Procedures
    Refrigeration at 40°F or below
    Potable water at 70°F or below
    Cooking
    Microwave
- Prevention of Cross-Contamination
    Cleaning and sanitizing (equipment, sinks, utensils)
    Preparation schedule or designated areas
    Product to product (such as chicken to fish)
- Personal Hygiene
    Illness Policy
    Hand washing
    Employee practices

PREPARATION

- Time–Temperature
    Batch preparation
    Time as a microbial barrier
    Prechill ingredients, equipment (salads and sandwiches)
- Use of Pasteurized Products
- Prevention of Cross-Contamination
    Cleaning and sanitizing
    Designated procedures
- Personal Hygiene
    Illness policy
    Hand washing
    Employee practices

COOKING

- Final Cooking Temperatures (Time)
    Beef, fish, eggs — 145°F (15 sec.)
    Pork, ground beef/pork, fish — 155°F (15 sec.)
    Poultry, stuffed meats — 165°F (15 sec.)
    Reheated foods — 165°F (15 sec.)
    Microwave — add 25°F to final cooking temperature

HOT HOLDING

- Time–Temperature
    140°F or above
- Use of Equipment
    Stir product
    Cover containers
    Utensil storage

- Breaking the Re-Run Cycle
- Batch Preparation

COOLING

- Time–Temperature
    140°F to 45°F or below in less than four hours
    New code
    140°F to 70°F within two hours and
    70°F to 40°F within four hours
- Procedures–Product Specific
- Procedures–Equipment Specific
- Prevention of Cross-Contamination
- Personal Hygiene
    Illness policy
    Hand washing
    Employee practices

REHEATING

- Time–Temperature
    Reheat to 165°F or above in less than two hours
- Prevention of Cross-Contamination
- Personal Hygiene
    Illness policy
    Hand washing
    Employee practices

SERVING

- Time–Temperature
    Hot foods at 140°F or higher
    Cold foods at 40°F or below
    Discard foods held above 40°F or below 140°F, after two hours
- Prevent Cross-Contamination
    Designated serving procedures
    Storage procedures at cook-at-service stations
    Self-service station procedures
- Personal Hygiene
    Illness policy
    Hand washing
    Employee practices

Incorporating Preventive Measures
in Recipes

When food safety and sanitation instructions are incorporated into a recipe, employees become familiar with the operation's standards and also learn what specific food safety criteria must be met.

The following are examples of such written instructions:

- Measure all temperatures with clean, sanitized thermometer or thermocouple.
- Wash hands before handling food.
- Wash hands after handling raw food.
- Wash hands after any interruption that may contaminate hands.
- Wash, rinse, and sanitize all equipment and utensils before and after use.
- Include in recipe procedures, if preparation is interrupted return all ingredients to refrigerated storage.
- Work in small batches or units at a time; refrigerate upon completion of each step.

In addition, a foodservice manager can:

- Establish in recipes time standards for completion of task units.
- Incorporate cross-contamination cautions for working with raw meat items.
- State precautions about foods that front-line employees may not realize are potentially hazardous, such as cooked vegetables, sprouts, tofu, sautéed onions, and garlic in oil.
- Include standards in regard to washing fruit and vegetables.
- List procedures and time standards for thawing of certain foods.
- Establish product handling standards at grill and prep-at-service stations; give specific handling instructions to prevent cross-contamination.
- Include reminders about prechilling salad ingredients.
- Review and update recipes periodically to determine whether a product or procedural change may reduce risk.

HACCP incorporates temperature and time controls into the recipe procedures, as well as operational procedures and preventive measures. It is also necessary to write specific details so that HACCP controls may be achieved.

For example, in a hamburger recipe the time/temperature control would be to cook the hamburger patties to 155°F and maintain that temperature for 15 seconds. A procedural control would be to check and record the hamburger patties and grill temperatures at specified times, such as during cooking of the first batch, just prior to peak service, after peak service, and during slow-service periods.

Review the Salmon Broccoli Casserole recipe on page 129. Answer the following three questions:

1. What time and temperature controls need to be written at the critical control points in the recipe food flow: cooking, hot holding, cooling, cold holding, and reheating?

2. What operational procedures would you write into the recipe? A production operational procedure might say something such as "Mix and bake 75 percent of the estimated daily production."

3. What special sanitation instructions would you write into the recipe that you expect your employees to follow? An example is, "Wash hands before handling food, after handling raw foods, and after any interruption that may contaminate hands." Employee hygiene and sanitation instructions will include much of the same wording from recipe to recipe, so you may want to include it in the same area or same manner in all your recipes.

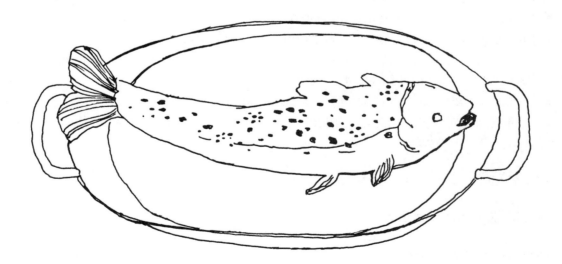

Salmon Broccoli Casserole

Items:

Onion
Oil
Mushrooms
Broccoli
Rice
Salmon, canned
Eggs, fresh
Cream sauce
Cheese

Pre-preparation:

Cook rice following rice recipe and handling instructions. Prepare cream sauce following the cream sauce recipe and handling instructions.

Preparation:

1. Sauté onions in oil.
2. Add mushrooms and broccoli, cook 1–2 minutes.
3. Combine vegetables with cooked rice, drained salmon, and eggs.
4. Stir cream sauce into salmon, rice, and vegetable mixture.
5. Top with cheese.
6. Bake in 350°F oven for 20 minutes until mixture is bubbly.

Answers are on page 130.

Salmon Broccoli Casserole

Items:

Onion
Oil
Mushrooms
Broccoli
Rice
Salmon, canned
Eggs, pasteurized whole
Cream sauce
Cheese

Pre-Preparation:

Cook rice following rice recipe and handling instructions. Prepare cream sauce following the cream sauce recipe and handling instructions.

Preparation:

1. Sauté onions in oil.
2. Wash vegetables, add mushrooms and broccoli, cook 1–2 minutes.
3. Combine vegetables with cooked rice, drained salmon, and eggs.
4. Stir cream sauce into salmon, rice, and vegetable mixture.
5. Top with cheese.
6. Bake in 350°F oven until internal temperature is 165°F (15 sec.).
7. Mix and bake 75 percent of estimated daily production.
8. Hold remaining cold ingredients separately at 40°F in refrigerator until needed.
9. Prepare in small batches as needed.

Holding:

10. Transfer immediately to a 150°F hot-holding box. Use within four hours.

Leftovers:

11. Unserved casserole should be rapidly cooled in an ice bath to 70°F within two hours and from 70°F to 40°F within four hours or less.
12. When cooled to 40°F, cover with film, store in refrigerator at 40°F.

Reheating:

13. Reheat to 165°F in less than two hours. Do not save a second time.

Special Instructions: • Wash hands before handling food, after handling raw foods, and after any interruption that may contaminate hands • Work with small amounts of products to maintain 40°F or below • Return all ingredients to refrigerated storage when preparation is interrupted • Measure temperatures with a clean, sanitized, and calibrated thermometer in the thickest or middle part of the product.

Rewriting a Recipe to Include Control Points

Rewrite the Mexican Chicken Mixture recipe below or to include HACCP controls for:

- Time and temperature
- Operational procedures
- Safe employee hygiene and sanitation procedures

Review the Salmon Broccoli Casserole recipe on page 130 as an example.

Mexican Chicken Mixture

Items:

Seasoned mix
Chicken meat, boneless, 1-inch cubes
Salad oil
Onions, chopped
Bell peppers, chopped
Potatoes, grated
Green olives, finely diced
Canned jalapeño pepper, seedless, finely chopped
Ground pear tomatoes
Water, ½ gal
Chicken seasoning mix

Preparation:

1. Place chicken on two sheet pans. Sprinkle with seasoning mix and bake in a hot oven for 20 minutes until done.
2. In a 24-quart sauce pot or kettle combine oil and onions and sauté until onions become translucent.
3. Add bell peppers, potatoes, green olives, jalapeño peppers, and ground tomatoes. Mix well.
4. Bring to a boil, and add chicken, chicken seasoning mix, and water. Mix well and simmer until potatoes are done.
5. Store chicken mix in 1½-gallon plastic containers. Place containers on cooling shelves for one hour. Label, date, and cover with clear film wrap. Make two small holes to allow steam to escape. Refrigerate until product reaches 40°F, then cover with lid.

Note: Be sure not to overcook chicken.

To ensure a good product and yield, make sure ovens are calibrated and in perfect working condition at all times.

Answers are on page 132.

Mexican Chicken Mixture

Items:

Seasoned mix
Chicken meat, boneless, 1-inch cubes
Salad oil
Onions, chopped
Bell peppers, chopped
Potatoes, grated

Green olives, finely diced
Canned jalapeño pepper, seedless, finely
  chopped
Ground pear tomatoes
Chicken seasoning mix
Ice

Preparation:

1. Place chicken on two sheet pans. Sprinkle with seasoning mix and bake in a hot oven for 20 minutes until done.
2. In a 24-quart sauce pot or kettle combine oil and onions and sauté until onions become translucent.
3. Add bell peppers, potatoes, green olives, jalapeño peppers, and ground tomatoes. Mix well.
4. Bring to a boil, and add chicken and chicken seasoning mix. Mix well and simmer until potatoes are done and **mixture reaches at least 165°F.**

Cooling:

5. **Cool mixture to 140°F, by stirring every 15 minutes. When mixture is at 140°F, add 1 gallon of ice to start rapid cooling. Stir in to melt.**
6. **Divide mixture into 2½ inch stainless steel hotel pans. Place pans in a bed of ice. When mixture cools to 40°F, remove from ice bath, cover with plastic film, and store in refrigerator at 40°F or below. Mixture should not be made more than 2 days in advance of use.**

   Note: **If using a prep sink for the ice bath, the prep sink needs to be washed and sanitized to prevent accidental cross-contamination with cooked products.**

Reheating:

Reheat in small batches, the amount needed for 2 to 4 hours of service, to 165°F in steamer, stirring every 10 minutes. Reheating should take less than 2 hours.

Holding:

Transfer immediately to a 150°F hot-holding box. Do not save leftovers.

Special Instructions: • Wash hands before handling food, after handling raw foods, and after any interruption that may contaminate hands • Wash and sanitize equipment in the dish machine between product changes • Maintain sanitizer buckets with fresh solutions; check concentration with test strip • Measure temperatures with a clean, sanitized, and calibrated thermometer in the thickest part of the product.

HACCP Project

In the following exercise, you will work through the third HACCP step.

1. Identify potentially hazardous foods (PHFs).
2. Identify critical control points.
3. **Establish control procedures to guarantee safe food.**
4. Establish monitoring procedures.
5. Establish corrective action.

Start by reviewing the Fried Chicken Breast recipe, then go to Step 3 on page 135.

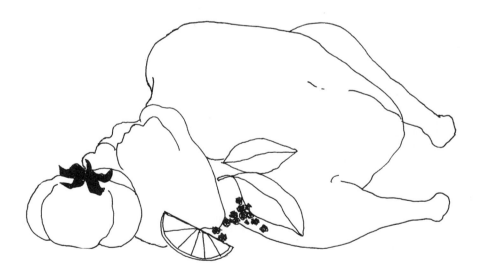

FRIED CHICKEN BREAST

Items:

Boneless chicken breast
Flour
Salt
White pepper
Eggs
Milk

Preparation:

1. *Mix* flour, salt, and pepper.
2. *Mix* egg with milk; blend well.
3. *Dip* chicken in egg wash, then in flour.
4. *Deep fat fry* in 375°F for approximately 4 minutes.
5. Remove, place on wire rack in sheet pan.
6. *Bake* in 375°F oven for approximately 20 minutes until done.
7. Remove and *chill.*
8. *Reheat* to serve.

HACCP Step 3: Establish Control Procedures to Guarantee Safe Food

FRIED CHICKEN BREAST

A. Part of recipe HACCP is to identify the controls that should be established for each process in the food flow. Using the Fried Chicken Breast recipe, **write a control procedure for each process in the food flow**.

- Receiving of PHF-chicken, egg, milk
- Storage of PHF
- Mixing and dipping procedure
- Handling procedure for fried and raw products (cross-contamination while frying)
- Holding after frying (time/temperature)
- Internal product temperature after baking
- Chilling time/temperature
- Reheating time/temperature
- Holding temperature until serving
- Employee hygiene and handling practices
- Potential for cross-contamination

Example:

1. All PHF received at 40°F or below in sound condition from approved source.

2.

3.

4.

5.

6.

7.

8.

9.

Answers are on page 136.

ANSWERS

HACCP Step 3: Establish Control Procedures to Guarantee Safe Food

FRIED CHICKEN BREAST

A. Part of recipe HACCP is to identify the controls that should be established for each process in the food flow. Using the Fried Chicken Breast recipe, **write a control procedure for each process in the food flow.**

- Receiving of PHF-Chicken, egg, milk
- Storage of PHF
- Mixing and dipping procedure
- Handling procedure for fried and raw products (cross-contamination while frying)
- Storage after frying (time/temperature)
- Internal product temperature after baking
- Chilling time/temperature
- Reheating time/temperature
- Holding temperature until serving
- Employee hygiene and handling practices
- Potential for cross-contamination

Example:

1. All PHF received at 40°F or below in sound condition from an approved source.
2. Store PHF at 40°F or below, store raw PHF separately and below other foods.
3. Egg wash and breading used in breading procedure is to be refrigerated or discarded after use.
4. Prevention of cross-contamination between raw and cooked chicken:
    - Use separate tongs for cooked and raw chicken.
    - Wash and sanitize all utensils and food contact surfaces in handling raw chicken.
5. Bake to internal temperature of 165°F or higher.
6. Cool to 70°F in less than two hours and from 70°F to 40°F in four hours or less.
7. Separate cooked product from raw during chilling.
8. Wash and sanitize thermometer between temperature checks.
9. Reheat to 165°F within two hours.
10. Hold chicken at 140°F or higher during service.
11. Employees follow good hand washing and hygiene practices.
12. Use clean, sanitized utensils and equipment.

HACCP Step 3

FRIED CHICKEN BREAST

B.   Rewrite the Fried Chicken Breast recipe to include HACCP controls for:

- Time and temperature
- Operational procedures
- Safe employee practices and sanitation procedures

For examples, review the Salmon Broccoli Casserole recipe on page 130 and the Mexican Chicken Mixture recipe on page 131.

Items:

Boneless chicken breast
Flour
Salt
White pepper
Eggs
Milk

Preparation:

Example:

1.   Mix flour, salt, and pepper.

2.   Mix egg with milk, blend well. Egg/milk mixture temperature should be 40°F or colder.
Egg wash is kept refrigerated or discarded after use.
Egg wash should not be used for any other product except raw chicken.

3.

4.

5.

6.

7.

8.

Answers are on page 138.

ANSWERS

HACCP Step 3

FRIED CHICKEN BREAST

B.   Rewrite the Fried Chicken Breast recipe to include HACCP controls for:

- Time and temperature
- Operational procedures
- Safe employee practices and sanitation procedures

For examples, review the Salmon Broccoli Casserole recipe on page 130 and the Mexican Chicken Mixture recipe on page 131.

Items:

Boneless chicken breast
Flour
Salt
White pepper
Eggs
Milk

Preparation:

Example:

1.   Mix flour, salt, and pepper.

2.   Mix egg with milk, blend well. Egg/milk mixture temperature should be 40°F or colder.

Egg wash is kept refrigerated or discarded after use.

Egg wash should not be used for any other product except raw chicken.

3.   Dip chicken in egg wash, then in flour.
Wash hands thoroughly before and after this procedure. The use of tongs or plastic gloves is recommended while working with the raw chicken.
Discard flour after using for chicken.

4.   Deep fat fry at 375°F for approximately 4 minutes. Complete cooking process by cooking chicken in oven to an internal temperature of 165°F.

Use different tongs for handling of raw and cooked chicken.

Wash and sanitize all utensils, knives, and cutting boards used in handling raw chicken.

5.   As chicken is removed from fryer, it should be placed in a single layer on a sheet pan. Bake in 375°F oven until chicken reaches an internal temperature of 165°F.
Check temperature at thickest part with a clean, sanitized thermometer.

6. Place chicken in a single layer on a sheet pan in cooler to cool quickly to 40°F. Chicken should cool to 40°F in 4 hours. Separate raw products from cooked products in storage.

7. Chicken should be reheated to 165°F in a 375°F oven. Heat chicken in a single layer on the pan, so it reheats quickly within 2 hours or less.

8. Hold chicken at internal temperature of 140°F in a heated holding cabinet. Destroy product after holding 2 hours.

Special Instructions: • Wash hands before handling food, after handling raw foods, and after any interruption that may contaminate hands • Wash and sanitize equipment in the the machine between product changes • Maintain sanitizer buckets with fresh solutions, check concentration with test strip • Work with small amounts of product to maintain 40°F or below • Work with only one tray of product at a time • Return all ingredients to refrigerated storage when preparation is interrupted • Measure temperatures with a clean, sanitized, and calibrated thermometer in the thickest part of the product.

> *Destiny is not a matter of chance; it is a matter of choice.*
>
> *– William Jennings Bryan*

STEP 4

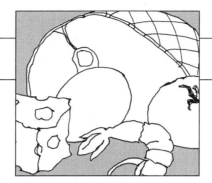

Establish Monitoring Procedures

Monitoring is a planned sequence of observations or measurements to assess whether a critical control point (CCP) is under control and to produce an accurate record for future verification. Examples of measurements for monitoring include:

Visual observations

Temperature

Time

pH

Water activity

What lies
behind us and
what lies
before us
are tiny matters
compared to what
lies within us.

—*Oliver Wendell Holmes*

APPLICATION OF STEP 4

Establish Monitoring Procedures

In each foodservice facility or kitchen, someone should be assigned the responsibility of monitoring the CCPs. If a product or process does not meet critical standards, immediate corrective action must be taken.

Example:

Assign one person to make and test sanitizer solutions each day.

Assign responsibility for equipment temperature logs.

Assign responsibility for food temperature logs during cooking, cooling, and reheating.

Flow chart the process identifying ingredients, time, temperature, equipment, and hand washing.

All records and documents with CCP monitoring must be signed and initialed by the person doing the monitoring.

Purposes of Monitoring

Monitoring serves three main purposes:

1. Monitoring is essential to food safety management, in that it tracks the system's operation. If monitoring indicates that there is a trend toward loss of control, then corrective action can be taken to bring the process back into control before a deviation or problem occurs.

2. Monitoring is used to indicate when there is loss of control and a deviation occurs at a CCP, that is, exceeding the critical limit.

3. Monitoring provides written documentation for use in verification of the HACCP plan.

Unsafe food may result if a process is not properly in control and a deviation occurs. Because of the potentially hazardous consequences of a critical defect, monitoring procedures must be effective. Continuous monitoring is possible with many types of physical, observational, and chemical methods:

Time/temperature logs
Food temperature monitoring charts
Equipment monitoring charts
Sanitizer checklists
Flow charts
Time/temperature graphs

Assignment of Monitoring

Assignment of responsibility for monitoring is an important consideration for each CCP. Selected individuals are often associated with production (line supervisors, certain line workers). Those individuals monitoring CCPs must:

- Be trained in the technique used to monitor each preventive measure;
- Fully understand the purpose and importance of monitoring;
- Have ready access to the monitoring activity;
- Be unbiased in monitoring and reporting; and
- Accurately report the monitoring activity.

Unusual occurrences must be reported immediately so that adjustments can be made in a timely manner to assure that the process remains under control. The person responsible for monitoring must also report a process or product that does not meet critical limits so that immediate corrective action can be taken.

Establishing Monitoring Procedures

The established criteria or standard for each CCP should be as specific as possible, such as: "Poultry must reach an internal temperature of 165°F or greater." The established standard provides clear directions in recipes and procedures. This allows the food handler to be aware of a situation in which a standard is not met.

Now that you have established criteria or standards for potential hazards, each standard must be easily monitored. Typical methods for monitoring CCPs may include:

Physical measurements (time and temperature logs).

Visual observations (watching worker practices, inspecting raw materials).

Sensory evaluations (smelling for off-odors, looking for off-colors, feeling for texture).

Chemical measurements (pH or acidity, water activity salt content; pH and water activity are discussed on pages 235, 237).

Once standards or critical limits are established for each CCP, the HACCP system must designate:

- The foods that must be monitored
- By which methods it should be monitored
- The person responsible for monitoring

Foodservice workers must record monitoring results for management review. Recordkeeping is an essential part of the HACCP system. These records indicate to management and regulatory agencies* that you have properly inspected, handled, and processed foods and ingredients.

Food Flows

Pages 147, 148, 149 give examples of simplified flow charts illustrating safe and unsafe food flows.

The Perilous Path assumes bacteria, viruses, and parasites may be found in food. If this food is thoroughly cooked, the result will be safe food. If the thoroughly cooked food is served immediately, it will be safe food. When the thoroughly cooked food is hot held at 140°F or higher, it will be safe food. If thoroughly cooked food is cooled rapidly and reheated rapidly to the proper temperature and within proper time limits, it will also be safe food. It is only through monitoring procedures at each critical control point that safe food can be determined.

---

*Regulatory agencies such as OSHA, the Joint Commission for Hospital Accreditation, and local health departments.

The More Perilous Paths illustrate the same safe food path with thorough cooking, but adds some unsafe paths. If food is not sufficiently cooked or served raw, it will be unsafe. If the food is thoroughly cooked, but any of the following factors are present—infected workers, dirty hands, contaminated equipment or utensils—you will have unsafe food.

The More Treacherous Routes show that there are some real concerns with cold, raw, or previously cooked foods. These include foods in salad bars and at self-service stations. Such foods will not receive cooking or heat treatment, and if contaminated, there is no "kill" step. You will observe that prevention of infected workers, dirty hands, and contact with contaminated equipment or utensils is the important control step with cold, raw, or previously cooked foods.

A plan that
does not
produce immediate
results is not
a plan,
it is the beginning
of failure.
—*Author Unknown*

The Perilous Path

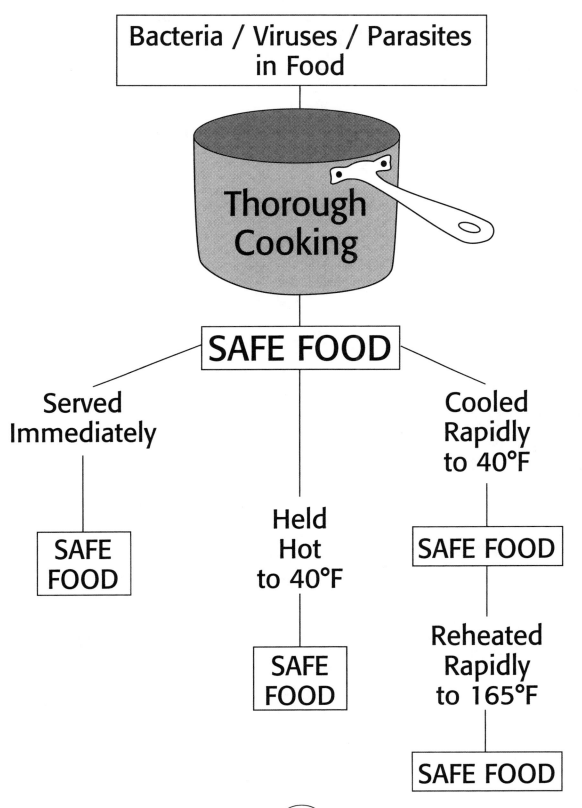

Bacteria / Viruses / Parasites in Food

Thorough Cooking

SAFE FOOD

Served Immediately

SAFE FOOD

Held Hot to 40°F

SAFE FOOD

Cooled Rapidly to 40°F

SAFE FOOD

Reheated Rapidly to 165°F

SAFE FOOD

More Perilous Paths

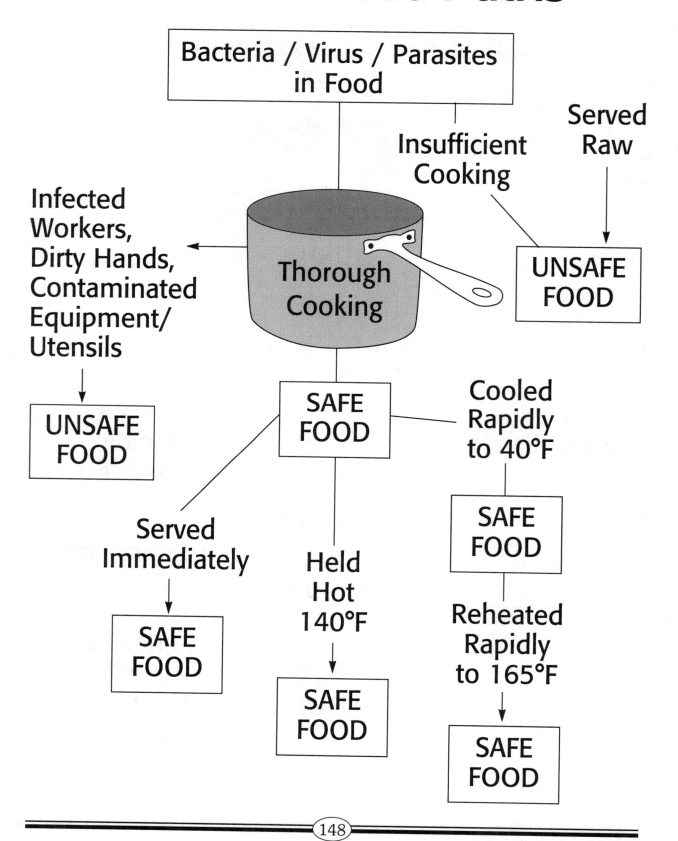

More Treacherous Routes

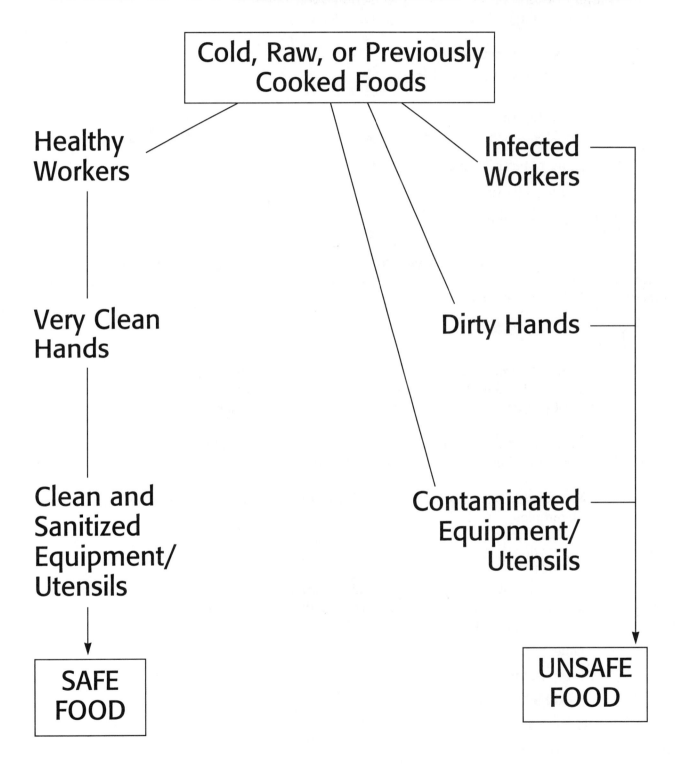

Charting Food Flows

The charting of food flows over time to identify food safety hazards is a crucial step in any HACCP system. First, start by assessing the risks in the following steps:

1. Analyze the menu
   Naturally contaminated foods
   Quantity preparation
   Multi-step preparation
   Cooking, cooling, reheating cycles

2. Identify potentially hazardous foods

3. Chart food flows

4. Observe procedures

Food preparation must be followed from beginning to end. This may be from delivery to service. The safety of some of the steps may be ascertained by asking questions, but most steps should be directly observed.

Steps

1. *Complete a menu analysis.* Look for potentially hazardous foods, foods that have been associated with outbreaks of illness in the past, complex foods requiring a great deal of preparation, and food produced in large quantities.

Start at a logical beginning of the process, such as with storage of the raw products or delivery. For some menu items, multiple ingredients may mean that there are several beginnings.

2. *Chart food flows on scratch paper.* Each step in the process gets a box and a name, for example:

   | mixing |

3. Make sure that you *observe the* critical *items.* Critical items are the problems and practices that are the leading factors in outbreaks of foodborne illness. Write down information at each step for:

   Time

   Temperature

   Food-handling practices

   Contamination sources

   Personal hygiene

   Any other relevant notes

For example:

> *Cooling*
>
> Pans not sanitized
> Stirred with hand
> Used covered metal pans
> Cooled in walk-in
> 170°F to 60°F in 17 hrs.

4. *Connect each operation with a flow arrow.* For example:

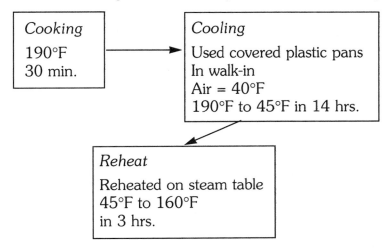

5. *Chart any loops or side operations.* For example:

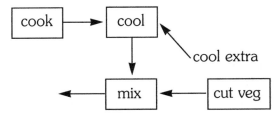

6. *Take continuous temperature observations* during cooking, cooling, reheating, and cold or hot holding. Graph relevant data.

Identifying Critical Control Points

Critical control points are the few steps in a process that, if not carried out correctly, may allow foodborne disease transmission. These points are operations or parts of operations where actual or potential hazards are found. They can be controlled to ensure that food safety is maintained by following these steps:

1. *Observe and chart food preparation flow.*

2. Identify those *steps in which contamination occurs* or may occur.

3. Identify those *steps that allow or may allow growth of microorganisms.*

4. Identify those steps that are *"kill" steps* or that correct for past problems.

5. Identify any steps in a process that present or may present a risk of injury or a *threat to the health of the consumer.* The following are examples:

    Cooking

    Cooling

    Points of cross-contamination

    Reheating

    Holding

6. Remember that it is best *not to rely on only one "layer" of protection.* For example, don't count on the cooking process to "kill" everything; cross-contamination can occur after cooking because of poor hygiene practices. Do not thaw raw seafood and chicken in the same container. The chicken can contaminate the seafood, since the seafood will be cooked only to 145°F and salmonella bacteria require 165°F to be destroyed. In each step of the preparation process, safe practices must be followed to reduce risk.

7. *Not all points are critical control points.* It is common practice to indicate that most or all points in a process are critical; however, this is not usually the case. If the concern is salmonella, then cooking will kill this organism and thus becomes a critical control point. This type of risk assessment rationale must be applied to each step in the food handling process.

See the example of identified control points and monitoring procedures for a potentially hazardous seafood salad on page 153.

Monitoring and Follow-Up

Once identified and established, critical control points must be monitored. The degree of monitoring and documentation is dependent on the type of establishment. Large, complex foodservice establishments may need to engage in extensive monitoring and record keeping to ensure the safe use of an HACCP plan. Smaller establishments, on the other hand, may need simple procedures that can be easily implemented.

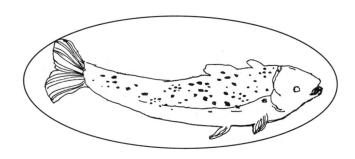

HACCP—Monitoring

PHF: Seafood Salad
(Potentially Hazardous Food)

CONTROL POINT	MONITORING PROCEDURES
Receiving	Inspection; measure/record temperatures
Pre-Chilling	Measure/record cooler air temperature every four hours
Mixing	Observe practices
Dishing	Observe practices
Cold Storage	Measure/record cooler air temperature every four hours
Replenishing	Observe practices
Displaying	Measure/record temperatures every two hours

FOOD FLOW CHART
Seafood Salad

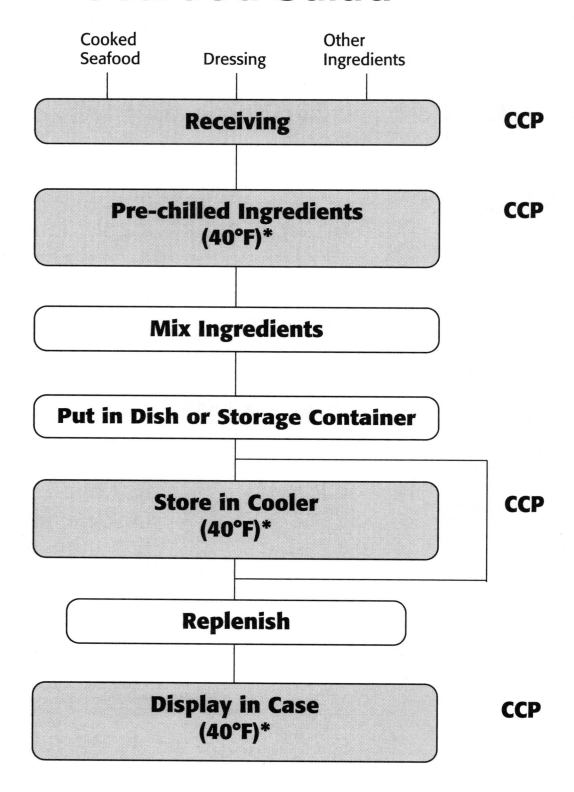

Cooked Seafood   Dressing   Other Ingredients

Receiving   **CCP**

Pre-chilled Ingredients (40°F)*   **CCP**

Mix Ingredients

Put in Dish or Storage Container

Store in Cooler (40°F)*   **CCP**

Replenish

Display in Case (40°F)*   **CCP**

*The Seafood industry has adopted a standard of 40°F or lower for seafood products.

Seafood Salad Flow Chart*

MAKE A FLOW CHART

To make a flow chart, you make a simple diagram showing the stages the salad will go through in preparation.

LIST INGREDIENTS

The ingredients of the salad can increase or decrease the risk of food safety problems. Raw animal foods contain spoilage and illness-causing bacteria. Most bacteria occur naturally in foods. Foodservice workers may add others during preparation.

Raw vegetables, such as celery and onions, contain spoilage and illness-causing bacteria. Thoroughly wash, or peel and wash, raw vegetables to remove any surface bacteria.

Some ingredients, such as lemon juice and mayonnaise, are acidic. Acidic ingredients may help to slow or stop bacterial growth.

MONITOR RECEIVING

Receiving is the first stage in the preparation. The safety of ingredients as they are received directly affects the safety of products prepared from those ingredients. Cooked seafood is more likely to have a problem with rapid bacterial growth than other ingredients in seafood salads.

Receiving is a critical control point (CCP) for seafood salad. Receiving is a stage at which potential hazards can be controlled. Moreover, later stages in the salad preparation process will not correct hazards. Therefore, receiving is a CCP.

When control problems occur at a CCP, such as receiving, immediate action is necessary. This means rejecting or discarding foods that do not meet the HACCP limits. For example, the following must be rejected:

- Damaged, spoiled, or contaminated ingredients
- Chilled cooked seafood with a temperature above 40°F
- Ingredients that do not meet company buying specifications

Record rejected or discard items on the invoice or on a Receiving Reject Form.

PRECHILL INGREDIENTS

As the salad is prepared, remember that rapid bacterial growth is still a potential hazard. Prechilling is a CCP because rapid bacterial growth can and must be controlled by prechilling the major ingredients to 40°F or below.

---

*Robert J. Price, Pamela D. Tom, and Kenneth E. Stevenson, *Ensuring Food Safety . . . The HACCP Way, A Resource Guide for Retail Deli Managers,* Extension Service, National Sea Grant Program, University of California, Cooperative Extension.

Monitor this CCP (prechilling) by measuring and recording the air temperature in the cooler every four hours. Record the temperature on a Cooler Temperature Form. Periodically, verify that the foods placed in the cooler are chilling rapidly by monitoring their temperature.

MIX SALAD INGREDIENTS

Mixing seafood with other ingredients can lead to bacterial and viral contamination. Contamination can come from worker's hands, utensils, or the mixer. This stage is not a CCP, because later stages in the process (storage and display at 40°F or below) will control any potential hazards.

However, even though this stage is not a CCP, you should not ignore it. To control this stage:

- Make sure the major ingredients are at 40°F or below.
- Avoid hand contact with the salad.
- Use clean utensils and mixer.
- Comply with sanitation and personal hygiene rules in your Standard Operating Procedures (SOPs).
- Follow good handling or preparation practices.

If necessary, modify the SOPs and handling practices to prevent contamination.

TRANSFER TO DISH OR CONTAINER

Transferring the seafood salad to a dish or storage container may result in contamination if the dish or container is not clean and sanitized. This step, too, is not a CCP, because later stages in the process (storage and display at 40°F or below) will control any potential hazards.

To prevent potential contamination:

- Use clean and sanitized dishes, containers, and utensils.
- Comply with sanitation and personal hygiene SOPs.
- Follow good handling or preparation practices.

If necessary, modify SOPs and handling practices to prevent contamination at this stage.

STORE IN SALAD COOLER

Bacterial growth continues to be a hazard during storage of the seafood salad in the cooler. Because it can and must be controlled during storage, this stage is a CCP. Set limits on the temperature and maximum storage time. For example:

- Set the cooler temperature below 40°F.
- Cover the container and label it with the date and time of preparation.
- Store the salad for only two to three days to ensure that bacterial growth is not a problem.

Temperature can be monitored by measuring and recording the air temperature of the cooler every four hours. Verify periodically that the cooler is keeping the salad at 40°F or below by measuring the temperature of the salad. If necessary, lower the cooler thermostat to keep the temperature of the salad at 40°F or below.

DISPLAY SALAD IN CASE

Growth of harmful bacteria in the cooler continues to be a potential hazard in the display case. This stage is a CCP. The limits of the display stage might include the following:

- Keep seafood salad at 40°F or below.
- Hold salad for no longer than two to three days.

Monitor the first limit by measuring and recording the temperature of the display case every four hours. Record the temperature on a Product Temperature Form. Verify periodically that the temperature of the product in the display case remains at 40°F or below.

RESTOCKING SALAD

When restocking the salad in the display case, remember that contamination can result from contact with workers' hands and utensils. This stage is not a CCP, because display at 40°F or below will control the potential hazard.

To control contamination:

- Transfer old salad to a smaller container; put the fresh salad in a new container.
- Do not add new salad on top of old salad.
- Use clean utensils and containers, and avoid hand contact with salad.
- Comply with sanitation and personal hygiene Standard Operating Procedures.
- Follow good handling or production practices.

Observe replenishing practices to make sure proper procedures are followed. If necessary, modify Standard Operating Procedures and handling practices to prevent contamination.

This example of setting up an HACCP plan for seafood salad shows how easy it is for most products. And an HACCP is a simple system for foodservice workers to follow. HACCP concentrates on critical hazards and will help prevent foodborne illness.

Customers may add potential safety problems, depending on how they handle and store the food. Instructions and informative labels may lower the probability of these safety problems.

SEAFOOD SALAD

Flow Chart	Potential Hazards	CCP	Critical Limits	Monitoring Procedures	Corrective Action
Cooked Seafood Dressing Other Ingredients					
Receiving	Rapid bacterial growth; contamination; foreign objects	CCP	Chilled items below 40°F; frozen items with no signs of thawing; no spoilage, contamination or foreign objects	Visual inspection; Measure/record temperature	Reject thawed frozen items, chilled items above 40°F, and items with spoilage, contamination, or foreign objects
Prechill Ingredients (40°F)	Rapid bacterial growth	CCP	Chill in shallow pans to below 40°F	Measure/record cooler air temperature every 4 hours	Adjust thermostat
Mix Ingredients	Contamination		Minimize hand contact; use clean utensils	Observe employee and handling practices	Modify Standard Operating Procedures (SOPs) and production practices
Put in Dish or Storage Container	Contamination		Use clean dish or container	Observe employee and handling practices	Modify Standard Operating Procedures (SOPs) and production practices
Store in Cooler 40°F	Rapid bacterial growth	CCP	Product below 40°F	Measure/record cooler air temperature every 4 hours	Adjust temperature; stir
Replenish	Contamination		Avoid hand contact	Observe employee and handling practices	Modify Standard Operating Procedures (SOPs) and production practices
Display in Case (40°F)	Rapid bacterial growth	CCP	Product below 40°F	Measure/record temperature every 2 hours	Adjust thermostat; stir

☐ = **Critical Control Point**

REFRIED BEANS FLOW CHART

(Example of Mishandled Food)

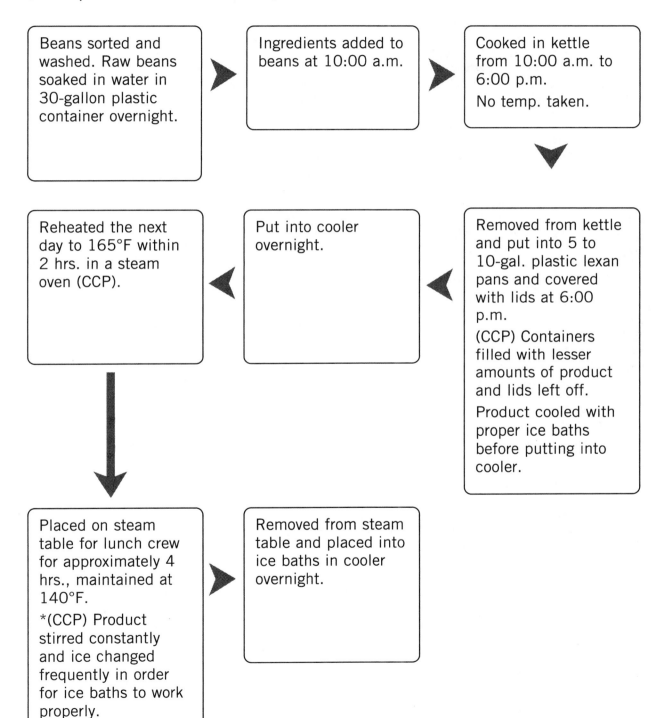

Beans sorted and washed. Raw beans soaked in water in 30-gallon plastic container overnight.

Ingredients added to beans at 10:00 a.m.

Cooked in kettle from 10:00 a.m. to 6:00 p.m.
No temp. taken.

Removed from kettle and put into 5 to 10-gal. plastic lexan pans and covered with lids at 6:00 p.m.
(CCP) Containers filled with lesser amounts of product and lids left off.
Product cooled with proper ice baths before putting into cooler.

Put into cooler overnight.

Reheated the next day to 165°F within 2 hrs. in a steam oven (CCP).

Placed on steam table for lunch crew for approximately 4 hrs., maintained at 140°F.
*(CCP) Product stirred constantly and ice changed frequently in order for ice baths to work properly.

Removed from steam table and placed into ice baths in cooler overnight.

*Eliminate risk: use dehydrated refried beans.

FLOW PROCESS FOR REFRIED BEANS

Ingredients: Pinto beans, lard, salt, bacon drippings, sodium benzoate, water

Preparation Step	Monitoring and Controls
COOKING Combine ingredients, add water, and cook in large stainless steel kettle (steam jacketed).	Assure cooking to temperature of at least 140°F with no interruption in the cooking process; monitor temperature during cooking with probe thermometer.
PREPARING Place cooked beans in stainless steel serving pans (mash to desired consistency).	Use clean, sanitized serving pans; avoid contamination during transfer; assure that depth of product in pans is no greater than 4 inches; use clean utensils to mash beans.
HOT HOLDING / SERVING Place serving pans into pass-through steamer for serving line supply.	Cook or serving line supervisor monitor internal temperature of product in pass-through unit and for serving line, with a calibrated probe thermometer. Assure pans remain at 140°F or above. Reheat if temperature falls below 140°F. Maintain temperature on serving line at 140°F. *OR*
COOLING Take pans to large walk-in cooler. Cool for storage for later use.	Using ice/water bath method, cool rapidly within 4 to 6 hours to 40°F. Monitor temperature during cooling. Reheat to 165°F, if not 40°F within 4 to 6 hours. Use 2-inch or less shallow pans to assure adequate cooling. Avoid stacking. Assure air flow to product.
COLD HOLDING Store pans in large walk-in cooler for use as needed.	Monitor internal temperature (use a calibrated probe thermometer) of product in walk-in. Maintain product at 40°F or below. Cover, label product with date and time it is cooled. Establish rotation system for finished product in walk-in. Use first in–first out rotation procedure. Avoid stacking. Assure air flow to product.
REHEATING Reheat for next days' use in steamer oven unit.	Reheat product rapidly to 165°F. Use probe thermometer to monitor adequate reheating. Establish time required to assure adequate reheating. Use timers.

ROTISSERIE CHICKEN FLOW CHART

Steps	Potential Hazard	Control/Solution/Monitoring
1. USDA plant source	Salmonella Campylobacter Factory contamination	No complete technological control at present time
2. Received chilled meat 32°F	Rapid bacterial growth Spoilage; contamination Foreign objects	Visual inspection Measure/record temperature
3. Transported and held in deli 38°F	Incomplete thawing can cause undercooking; Rapid bacterial growth	Observe thawing Modify thawing practices
4. Rinsed in 3-compartment sink	Cross-contamination	Designated equipment Monitor equipment cleaning Sanitizing
5. Seasoned	Cross-contamination Unauthorized ingredients (Nitrates/MSG/colors)	Observe practices; Modify practices Approved ingredient control
6. Skewer/Whole deli walk-in	Cross-contamination	Designated storage and equipment Modify practices
7. Roasted 90 min. 450°F	Pathogen survival with incomplete cooking	Follow time/temperature instructions Measure/record center temperature Continue cooking until center reaches 165°F
8. Skewer rotated during roasting	Pathogen introduction (drip raw to cooked; contaminated equipment such as gloves)	Designated equipment Modify practices
9. Removed from skewer/package	Contamination	Designated equipment Modify practices Sanitation
10. Hot held 140°F (30 min-2 hrs. est.)	Rapid bacterial growth	Monitor temperature; measure/record center temperature every 2 hours Limit time held
11. Leftover tray cool down; wrap and display	Rapid bacterial growth Pathogen introduction Pathogen growth	Avoid hand contact Cool down quickly with small mass amounts Monitor/record product temperature Stock rotation Manage inventory to preclude leftovers

Customer handling after sale

GREEN CHILI FLOW CHART*

(Example of Mishandled Product)

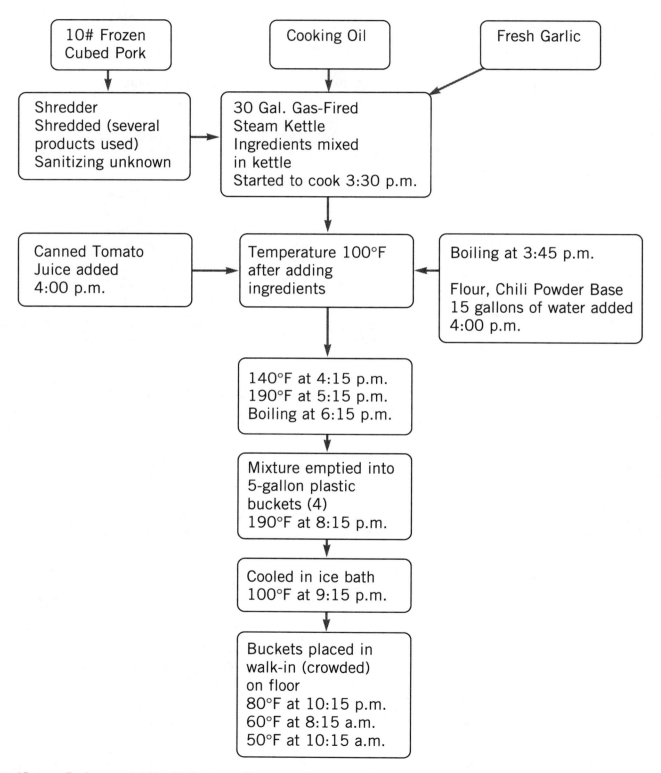

10# Frozen
Cubed Pork

Cooking Oil

Fresh Garlic

Shredder
Shredded (several
products used)
Sanitizing unknown

30 Gal. Gas-Fired
Steam Kettle
Ingredients mixed
in kettle
Started to cook 3:30 p.m.

Canned Tomato
Juice added
4:00 p.m.

Temperature 100°F
after adding
ingredients

Boiling at 3:45 p.m.

Flour, Chili Powder Base
15 gallons of water added
4:00 p.m.

140°F at 4:15 p.m.
190°F at 5:15 p.m.
Boiling at 6:15 p.m.

Mixture emptied into
5-gallon plastic
buckets (4)
190°F at 8:15 p.m.

Cooled in ice bath
100°F at 9:15 p.m.

Buckets placed in
walk-in (crowded)
on floor
80°F at 10:15 p.m.
60°F at 8:15 a.m.
50°F at 10:15 a.m.

*Denver Environmental Health Services, Consumer Protection/Food Safety.

In each foodservice operation, someone should be assigned the responsibility of monitoring the CCPs. Immediate corrective actions must be taken for any product or process that does not meet critical limits.

Flow Chart Quiz

1. Review the flow chart on page 162. What problems do you note in the processing of the green chili?

2. Where are the critical control points (CCPs) in the flow chart?

3. What immediate corrective action needs to be taken?

4. What recommendation would you make to reduce the risk of foodborne illness?

5. How would you establish monitoring procedures?

6. Who should be responsible for monitoring the CCP?

7. Complete a Time/Temperature Graph for the green chili.

Answers are on pages 165, 166.

ANSWERS

FLOW CHART QUIZ

1. Review the flow chart on page 162. What problems do you note in the processing of the green chili?

    Cool-down procedure. In the green chili flow chart, you will notice that after 13 hours of cooling the chili has not cooled down to 40°F. To cool in the required time of 4 to 6 hours, the product must be divided into smaller units, preferably metal containers, and cooled in ice baths with stirring until the product is at 40°F. After cooling to 40°F the product must be placed in the refrigerator, or after cooling to 70°F the product may be transferred to shallow 3-inch pans and placed in the refrigerator with good air circulation. The pans must not be stacked until the product reaches 40°F.

2. Where are the critical control points (CCPS) in the flow chart?

    Shredding. Raw pork could contaminate other products if not properly cleaned and sanitized.
    Cooling. Product was not cooled to 40°F in six hours.
    Temperature. Product was in the danger zone more than four hours.

3. What immediate corrective action needs to be taken?

    Product should be discarded. With risk of cross-contamination and product in danger zone more than four hours (over thirteen hours in this situation), there have been plenty of favorable time and conditions for bacterial growth, which produces a high risk for foodborne illness.

4. What recommendation would you make to reduce the risk of foodborne illness?

    Incorporate control procedures into recipes. In this situation, to speed cool down, ice could be substituted for part of the water and added after product reaches 140°F for a very fast cool down.
    Train employees in the proper cleaning and use of sanitizers. Enforce adherence to these procedures, as well as proper and frequent hand washing. Take temperatures of product at critical times, record temperatures on chart, and file charts for future reference and records. Incorporate control procedures into recipes. Then check that procedures are being followed and monitored.

5. How would you establish monitoring procedures?

    Time/temperature logs should be kept, and all products prepared in advance.
    All other products prepared for serving should have temperatures recorded after heating and during holding, and after cool down if product is to be reheated.
    Maintain equipment temperature logs, delivery temperatures and sanitizer (ppm) logs.
    Complete periodic flow charts and time/temperature graphs. This is also a good way to actually show and teach staff what is happening.

6. Who should be responsible for monitoring the CCPs?

    The manager, supervisor, and chef should be responsible.
    Foodservice workers are responsible for their assigned parts.

7. Complete a Time/Temperature Graph for the green chili.

Time/Temperature Graph

Date: _____

Product: _____

Container: _____

Method of Cooling: _____

Temp:

165°	
160°	
155°	
150°	
145°	
140°	
135°	
130°	
125°	
120°	
115°	
110°	
105°	
100°	
95°	
90°	
85°	
80°	
75°	
70°	
65°	
60°	
55°	
50°	
45°	
40°	

Time (hours)

Time/Temperature Graph

Date: _____

Product: Green Chili with Pork

Container: 5 gal plastic buckets

Method of Cooling: ice bath

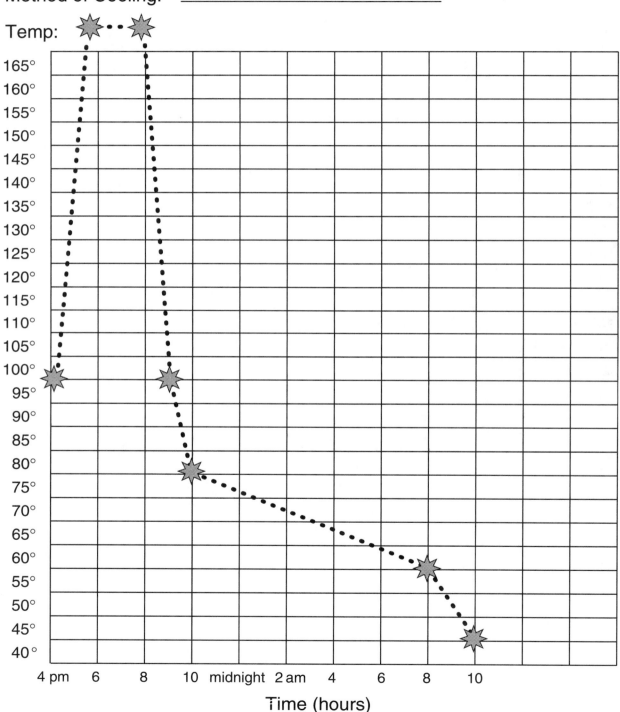

Time (hours)

HACCP Project

In the following exercise, you will work through the fourth of the HACCP steps.

1. Identify potentially hazardous foods (PHF).
2. Identify critical control points.
3. Establish control procedures to guarantee safe food.

4. Establish monitoring procedures.

5. Establish corrective action.

Start by reviewing the Fried Chicken Breast recipe, then go to Step 4.

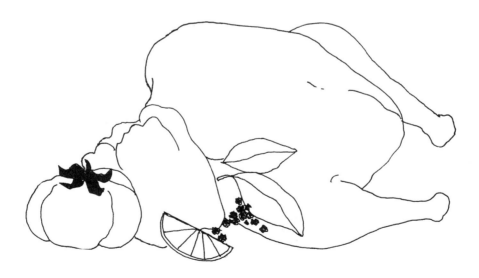

FRIED CHICKEN BREAST

Items:

Boneless chicken breast
Flour
Salt
White pepper
Eggs
Milk

Preparation:

1. *Mix* flour, salt, and pepper.
2. *Mix* egg with milk; blend well.
3. *Dip* chicken in egg wash, then in flour.
4. *Deep fat fry* in 375°F for approximately 4 minutes.
5. Remove, place on wire rack in sheet pan.
6. *Bake* in 375°F oven for approximately 20 minutes until done.
7. Remove and *chill.*
8. *Reheat* to serve.

HACCP Step 4: Establish Monitoring Procedures
FRIED CHICKEN BREAST

A. Write procedures to be monitored.

1. Internal Temperature

2. Chilled

3. Handling

4. Reheating

B. If this were your kitchen, write how these procedures are to be monitored, by whom, how often, and how they are to be documented.

Answers are on page 170.

ANSWERS:

HACCP Step 4: Establish Monitoring Procedures
FRIED CHICKEN BREAST

A. Write procedures to be monitored.

   1. Internal Temperature
       Check internal product temperature of chicken after baking—must be 165°F.
   2. Chilled
       Chicken must be chilled to an internal temperature of 40°F or below within 4 to 6 hours or less.
   3. Handling
       Chicken must be handled and stored in a manner preventing cross-contamination.
   4. Reheating
       Chicken must be reheated to 165°F within 2 hours.

B. If this were your kitchen, write how these procedures are to be monitored, by whom, how often and how they are to be documents.

   - *Manager, owner, or food service director* must first be committed to food safety matters and food safety leadership. He or she must demonstrate commitment through training, food safety committees, and incentives.

   - *Supervisors and managers* should be trained in food safety matters, food safety leadership, and coaching, and empowered to take corrective action at any time to prevent a problem. Supervisors should be responsible for making sure that food is handled and stored in a manner to prevent cross-contamination and that assigned temperature logs are completed by foodservice workers.

   - *Foodservice workers* are to be trained in food safety matters and provided with training prior to all new assignments on specific hazard controls. Training should be updated at least annually or as work processes and ingredients change. Foodservice workers should be assigned responsibility to take and record temperatures at critical control points in the preparation, serving, holding, cool down, and storage of PHFs. Workers should be assigned responsibility to record equipment temperatures at designated times.

HACCP Step 4: Flow Charting
FRIED CHICKEN BREAST

Mishandled Product

C. In the diagram on page 172, flow chart the Fried Chicken Breast, using the following observations. Mark the CCPs.

- Chicken is received as an individually quick frozen (IQF) fillet. Product temperature 0°F or below.

- At 8:00 p.m. you arrive at the restaurant. The chicken is found thawing in the refrigerator. Chicken temperature 38°F.

- The chicken is removed from the refrigerator, dipped in egg/milk mixture before being floured.

- At 9:00 p.m. the chicken is cooked in oil; internal product temperature is 140°F.

- The employee preparing the chicken is using the same tongs to place raw chicken in the fryer and to place the cooked chicken in a pan.

- At 9:30 p.m. the chicken is placed in the warmer to sell; internal temperature is 135°F.

- At 11:00 p.m. the chicken is placed in cooler to store overnight. Leftover product is placed in 2-inch deep pans, covered with foil. Chicken temperature 120°F. Containers stacked three high in refrigerator. A recording thermometer verifies that after 8 hours the product cooled to only 65°F.

Next Day

- At 7:00 a.m. leftover chicken is reheated by placing it in the warmer.

- At 10:00 a.m. reheated temperature reaches only 112°F in food warmer when it is being sold to customers. Product is coded for discard after two hours holding in warmer.

FOOD FLOW CHART

Fried Chicken Breast

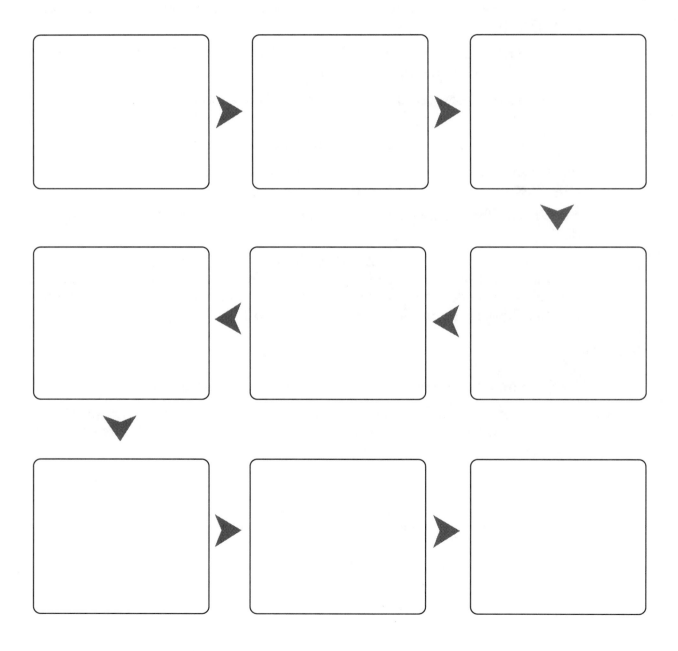

Answers are on page 174.

ANSWERS:

FOOD FLOW CHART

Fried Chicken Breast

IQF frozen chicken breast fillet 0°F

Pulled chicken to thaw for 24 hours in refrigerator in original carton on sheet pan on bottom shelf, previous day.

Chicken removed from refrigerator 38°F. Dipped in egg/milk mixture, then floured.

8:00 p.m.

Chicken is placed in warmer to sell, 135°F. Leftover product used to start next day.

10:00 p.m.   CCP

Same pair of tongs is used for placing raw chicken in fryer and placing cooked chicken in pans.

CCP

Chicken is cooked in fryer; 140°F when removed from fryer.

9:00 p.m.   CCP

Placed in 2"-deep pans, covered with foil, stacked in 3 pans high in refrigerator. Temperature 120°F.

11:00 p.m.   CCP

Temperature 65°F. Reheated in warmer for 3 hours.

7:00 a.m.   CCP

Serving temperature 112°F when sold to customer. Product coded for discard after 2 hrs.

10:00 a.m.   CCP

D.   Graph temperatures recorded from findings and flow chart on page 174.

Time/Temperature Graph

Date: _____

Product: _____

Container: _____

Method of Cooling: _____

Temp:

170°	
165°	
160°	
155°	
150°	
145°	
140°	
135°	
130°	
125°	
120°	
115°	
110°	
105°	
100°	
95°	
90°	
85°	
80°	
75°	
70°	
65°	
60°	
55°	
50°	
45°	
40°	

Time (hours)

Answers are on page 176.

HACCP Step 4: Time/Temperature Graph

D.   Graph temperatures recorded from findings and flow chart on page 174.

Time/Temperature Graph

Date: _____

Product: ___Fried Chicken Breast___

Container: _____

Method of Cooling: ___Stacked, covered pans___

Temp:

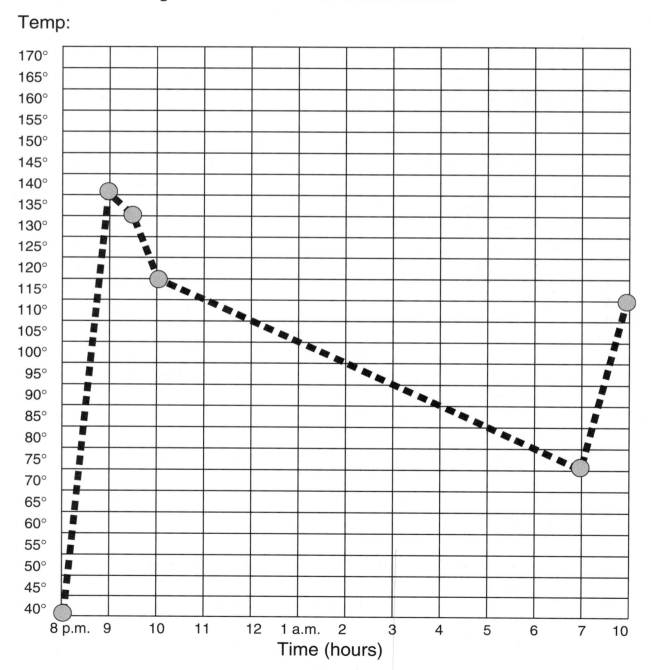

STEP 5

Establish Corrective Action

The HACCP system of food safety management is designed to identify potential health hazards and to establish strategies to prevent their occurrence. However, ideal circumstances do not always exist. When the unexpected occurs, corrective action plans must be in place to:

Determine whether a food should be discarded

Fix or correct the cause of the problem

Maintain records of corrective action taken

*S*uccess on any
major scale
requires you
to accept
responsibility. . . .
In the final
analysis, the one
quality that all
successful
people
have . . . is
the ability to take
on responsibility.

–Michael Korda

APPLICATION OF STEP 5

Establish Corrective Action

Actions must demonstrate that the CCP has been brought under control. Individuals who have a thorough understanding of the HACCP process, product, and plan are to be assigned the responsibility for taking corrective action. Corrective action procedures are to be documented in the HACCP plan.

Develop procedures for:

Problem reporting

Problem investigation

Corrective action

Follow-up

Procedures to Handle Critical Control Problems*

Problems occur when HACCP limits are not met. You will need to set up procedures to deal immediately with such failures. These procedures are called *corrective actions*. Examples of corrective actions include:

- Rejecting products not meeting buying specifications
- Adjusting a cooler's thermostats for proper temperature control
- Extending a product's cooking time
- Recooking or reheating a product to the proper temperature
- Modifying food handling procedures
- Destroying product that has risk of contamination or has been too long in the danger zone

Corrective Actions**

A corresponding corrective action must be established for each standard.

- Reject product
- Evaluate product
    - Adjust temperature
    - Move product
    - Cover product
- Evaluate procedure
    - Wash and rinse
    - Clean and sanitize
    - Redo
- Discard product

---

*Ensuring Food Safety—The HACCP Way: An Introduction and Resource Guide for Retail Deli Managers, University of California Cooperative Extension, Davis, California.

**Adapted with permission from *Managing a Food Safety System* seminar. Copyright © 1992 by the Education Foundation of the National Restaurant Association.

The following table lists the corrective actions that may be established for various control points.

Control Point	Corrective Action
Source	If a product is not from an approved, inspected source, reject the product and refuse to order from that source.
Delivery	If a product is not fresh or does not meet specifications, refuse the product and send it back.
Cooking	If a product is not cooked to the recommended product temperature, pull the product from service and continue cooking until the proper temperature is reached.
Handling	If cross-contamination occurs with a cooked product, destroy the contaminated product.
	If poor personal hygiene and food handling practices are occurring, correct the problem, update training, and enforce standards. If the problem can be related to an immediate situation and product, the product should be discarded.
Holding	Hold for service at 140°F or higher. If a product is not held above 140°F or below 40°F, correct the holding temperature. If a product has been in the danger zone more than four hours, discard it.
Cooling	Cool to 40°F or cooler in four-six hours or less. If a product does not reach 40°F in four-six hours, discard the product.
Cold Storage	Store cold foods at 40°F or cooler. If a product has been in the danger zone (above 40°F or below 140°F) for more than four hours, discard it.
Reheating	Reheat product to 165°F. If a product does not reach 165°F in two hours or less, discard it. Never mix new product with old.
Serving	If a product has not been handled properly at any of the preceding control steps, do not serve—discard.

The following are three examples from a deli operation. They show how controls have been written into the deli procedures. These procedures tell employees when to discard the product, based on the company standards for time.

These examples do not address corrective action based on temperature abuse. Employees must be trained in corrective action procedures and in knowing when to bring a problem to management's attention if they think that a product should be discarded. Corrective action must be written into the operational procedures; for example, "Discard a cold meat product that has been in the temperature danger zone above 45°F or higher for two hours or longer."

Fried Chicken — Hot Case

Received fresh, packed in ice   35°F maximum

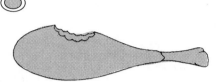

Held in cooler at   38°F maximum

On load days, check internal temperature of truck.

Record temperature on the distribution invoice.

One case is brought out of cooler, breaded, and placed back in cooler (following breading procedure)

No more than 5 minutes is taken to bread each case.

★ *Using clean, gloved hands*, place chicken in preheated 325°F fryer.

Make sure to dislocate thighs, to ensure thorough cooking.

Close lid, simmer for 14 minutes, turn timer on.   **ON**

When 14-minute cycle is finished, turn timer off. Wait for pressure to release, then slowly open fryer.

★ *Using clean, gloved hands,* use insulated gloves to remove basket from inside fryer. Place chicken onto pan.

★ Perform probe test. *Using clean, gloved hands*, clean thermometer with alcohol swab. Insert thermometer into large piece of chicken. Temperature should be 165°F minimum.

Log temperature on daily temperature chart.

Place chicken into preheated 150°F hot case. Discard after four hours if not sold.

Sell

★ Temperature tests at 9:00 a.m., 11:00 a.m., 1:00 p.m., 3:00 p.m., 5:00 p.m., 7:00 p.m. Holding temperature at 150°F minimum.

Repeat process throughout day as needed.

8:00 p.m. Discount remaining chicken, place in bunker, no earlier than 8:30 p.m.

★ *Discard remaining chicken* when store opens at 7:00 a.m. the next morning.

★ = Critical control point

Lunch Meats and Cheeses

Receiving temperature   35°F maximum

Held in cooler at   38°F maximum

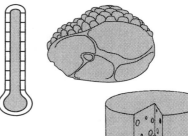

On load days, check internal temperature of truck.

Record temperature on the distribution invoice.

★ Daily temperatures should be taken

at 9:00 a.m., 11:00 a.m., 1:00 p.m., 3:00 p.m., 5:00 p.m., 7:00 p.m.

Log temperatures in Temperature Log Book.

Pull old product from case, put new product in, and replace old product on top.

Make sure not to place product up against the glass, so as not to block the air flow.

Keep product as least 8 inches below lights to maintain temperature below 38°F.

★ When slicing product, use clean, gloved hands. Cut open top package, place on clean slicer.

After slicing, wrap package tightly, immediately put back in case.

Sell

After 10 days, discard out-of-date product.

★ = Critical control point

Poor Boy Sandwiches

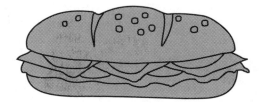

Using ends and over slices of meats and cheese.

Keep in separate airtight containers.

⭐ *Using clean and gloved hands,* assemble sandwiches, taking no longer than 30 minutes (1 minute per sandwich).

Wrap, price, and code date (one day only *maximum*) each sandwich.

Place in bunker case.

Sell

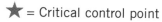

Discard on following business day.

⭐ = Critical control point

HACCP Project

In the following exercise, you will work through the fifth HACCP step.

1. Identify potentially hazardous foods (PHFs).

2. Identify critical control points.

3. Establish control procedures to guarantee safe food.

4. Establish monitoring procedures.

5. **Establish corrective action.**

Start by reviewing the Fried Chicken Breast recipe, then go to Step 5.

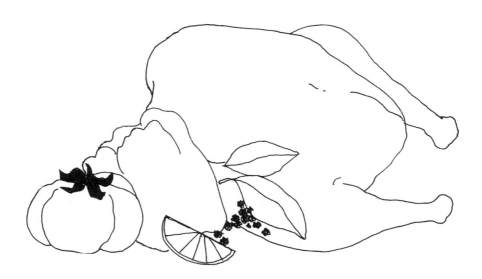

FRIED CHICKEN BREAST

Items:

Boneless chicken breast
Flour
Salt
White pepper
Eggs
Milk

Preparation:

1. *Mix* flour, salt, and pepper.
2. *Mix* egg with milk, blend well.
3. *Dip* chicken in egg wash, then in flour.
4. *Deep fat fry* in 375°F for approximately 4 minutes.
5. Remove, place on wire rack in sheet pan.
6. *Bake* in 375°F oven for approximately 20 minutes until done.
7. Remove and *chill*.
8. *Reheat* to serve.

HACCP Step 5: Establish Corrective Action
FRIED CHICKEN BREAST

A. When there is a breakdown in the control standards, how is the product to be handled? Listed below are possible breakdown points. Write a corrective action at each control point. Review the table on page 181.

Control Point	Corrective Action
Source	If product is not from an approved inspected source, reject product and/or refuse to order from that source.
Delivery	
Cooking	
Handling	
Chilling	
Reheating	
Serving	

Answers are on page 188.

HACCP Step 5: Establish Corrective Action

B.   Review the Fried Chicken Breast procedures, flow chart, and time/temperature graph on pages 171–176.

    1. Identify the steps in which the product was mishandled.

    2. Write a corrective action for each mishandled step.

C.   After reviewing these HACCP steps, *list three changes* you are going to implement in your operation tomorrow.

    1.

    2.

    3.

D.   List the three new ideas that made the biggest impression on you to bring about change in the way things are currently being done in your operation.

    1.

    2.

    3.

E.   Think of your menu. What do you believe to be the three menu items that have potential for the highest risk? What can you do to reduce the risk of these menu items?

    1.

    2.

    3.

F.   On what subjects do you need additional information?

ANSWERS:

HACCP Step 5: Establish Corrective Action

FRIED CHICKEN BREAST

A. When there is a breakdown in the control standards, how is the product to be handled? Listed below are possible breakdown points. Write a corrective action at each control point. Review the table on page 181.

Control Point	*Corrective Action*
Source	If a product is not from an approved, inspected source, reject the product and refuse to order from that source.
Delivery	If a product is not fresh or does not meet specifications, and temperature guidelines, refuse product and send back.
Cooking	If chicken is not cooked to an internal temperature of 165°F or higher, pull product and recook product until proper temperature is reached.
Handling	If cross-contamination occurs with cooked product, destroy contaminated product.
	If poor personal hygiene and food handling practices are occurring, correct problem, update training, and enforce standards. If problem can be related to an immediate situation and product, product should be destroyed.
Chilling	If product does not reach 40°F in six hours or less, discard product.
Reheating	If product does not reach 165°F in two hours or less, discard.
Serving	If product has not been handled properly at any of the preceding control steps, *do not serve—destroy.*

B. Review the Fried Chicken Breast procedures, flow chart and time/temperature graph on pages 171–176.

1. Identify the steps in which the product was mishandled.

2. Write a corrective action for each mishandled step.

● At 9:00 P.M. the chicken is cooked in oil. The product's internal temperature is 140°F.

Corrective action: Place chicken in a single layer on a sheet pan, continue cooking in oven to internal temperature of 165°F.

- The employee preparing the chicken is using the same tongs to place raw chicken in the fryer and to place cooked chicken in the pan.

  Corrective action: Recook the contaminated chicken to an internal product temperature of 165°F.

  Change procedures to include using separate tongs for raw product and cooked products.

- At 9:30 P.M. the chicken is placed in warmer to sell. Internal product temperature is 135°F.

  Corrective action: Pull chicken from warmer and serving line, place chicken in single layer on sheet pans, and continue cooking until internal product temperature of 165°F is reached.

- At 11:00 P.M. the chicken is placed in the cooler to store overnight. Leftover product is placed in two-inch deep pans, covered with foil. Chicken temperature is 120°F. Containers are stacked three high in refrigerator.

  Corrective action: Place leftover chicken in a single layer on a sheet pan. Place chicken on upper shelf of walk-in refrigerator or place sheet pans in an open rack with a minimum of four inches of space between sheet pans. Do not cover sheet pans until chicken has cooled to 40°F.

- At 7:00 A.M. leftover chicken is reheated by placing it in the warmer.

  Corrective action: Discard the chicken, since it was improperly cooked and cooled and was in the danger zone for more than four hours.

  If the chicken had been properly cooked and cooled, the chicken could be rapidly reheated in an oven to 165°F within two hours and then hot held at 140°F.

- At 10:00 A.M. reheated temperature is only 112°F in the food warmer.

  Corrective action: Discard the chicken. It should have been discarded in previous step.

  Change procedures for reheating; a warmer is not designed for rapid reheating.

  After reheating to 165°F within two hours, a product may be held at 140°F for two hours before discarding it.

STEP 6

Establish Effective Record-Keeping Procedures

Use record keeping and routine reviews of records to make sure that controls work.

A daily record review ensures that controls are working—that proper information was recorded and that workers handled foods properly. If records indicate potential problems, investigate immediately.

The associated records that document the HACCP system must be on file at the food establishment.

*I*f you have
built castles
in the air,
your work need
not be lost;
that is where
they should be.
Now put
the foundations
under them.

–Henry David Thoreau

Establish Effective Record-Keeping Procedures

Generally the total HACCP System Plan will include:

Listing of the HACCP team and assigned responsibilities

Description of the food and its intended use/product description/specifications

Listing of all regulations that must be met

Temperature monitoring logs

Copies of flow charts from receiving to consumption

Hazard assessment at each step in flow diagrams

Critical limits will be established for each hazard variable at each step:

Management

Employees

Customers

Environment

Facility

Equipment

Materials and Supplies

Food Production Methods—Handling from Source to Consumption

Corrective Action Plans

Procedures for Verification of HACCP System

Record Keeping*

Documentation is needed to record measurements that show standards are being monitored. These are the components of an effective HACCP record-keeping plan.

- Time/temperature logs and graphs
- Checklists
- Forms
- Flow charts
- Corrective actions
- Employee training records
- Product specifications

Examples of Monitoring Forms

In the HACCP system, documentation must be kept throughout the operational food flow. Using monitoring forms at receiving, preparation, cooking, cooling, reheating, and storage stages provides a product history, verifying that a product meets standards or indicating when adjustments to the system are needed.

Equipment temperatures should be monitored every four hours: this includes all refrigeration, cooking, and holding equipment. If necessary, adjust the thermostats so products meet the required temperature standards.

Other types of documentation concern standards of operational procedures (SOPs), including sanitation practices, employee practices, and employee training. This may be an informal notation of observations of what is working well and what is not working well. This documentation identifies practices and procedures that may have to be modified, as well as training needs.

The following monitoring forms may be adapted for your foodservice operation.

---

*Adapted with permission from *Managing a Food Safety System* seminar. Copyright © 1992 by the Education Foundation of the National Restaurant Association.

Receiving Temperature Form

Date	Product	Temperature	Initials/Comments

Reviewed by: _____   Date: _____

Receiving Reject Form

Date	Product	Rejected for:	Initials/Comments

Reviewed by: _____   Date: _____

Cooler Temperature Form

Date:	6:00 a.m. Temp.	10:00 a.m. Temp.	2:00 p.m. Temp.	6:00 p.m. Temp.	10:00 p.m. Temp.	Initials/Comments

Reviewed by: _____   Date: _____

Time/Temperature Log

Date: **Food Item:**   **Taken by:** _____

	time					
	temp					

	time					
	temp					

	time					
	temp					

	time					
	temp					

	time					
	temp					

	time					
	temp					

	time					
	temp					

	time					
	temp					

Display Product Temperature Form

Product:	Product Temperatures					Initials
Time:						
7:00 a.m.						
9:00 a.m.						
11:00 a.m.						
1:00 p.m.						
3:00 p.m.						
5:00 p.m.						
7:00 p.m.						
9:00 p.m.						
11:00 p.m.						
1:00 a.m.						
3:00 a.m.						
5:00 a.m.						

Reviewed by: _____   Date: _____

Food Temperature Monitoring Chart

Week Ending:

Menu Items:	Saturday		Sunday		Monday		Tuesday		Wednes.		Thursday		Friday	
	a.m.	p.m.	a.m.	p.m.	a.m.	p.m.	a.m.	p.m.	a.m.	p.m.	a.m.	p.m.	a.m.	p.m.

Equipment Monitoring Chart

Week Ending:

Menu Items:	Saturday		Sunday		Monday		Tuesday		Wednes.		Thursday		Friday	
	a.m.	p.m.	a.m.	p.m.	a.m.	p.m.	a.m.	p.m.	a.m.	p.m.	a.m.	p.m.	a.m.	p.m.
Freezer #1														
#2														
Freezer #1														
#2														
Products														
#1														
#2														
#3														
Deliveries														
#1														
#2														
#3														
Sanitizer (ppm)														
#1														
#2														
#3														
#4														

Sanitizer Checklist

Date: _____   Taken by: _____

Comments

Sanitizer in Use   Y/N   _____

    Quat 200 ppm   ppm   _____

    Chlorine 100–200 ppm   ppm   _____

    Wiping cloths sanitized   Y/N   _____

Equipment Sanitized   _____

    Cutting boards   Y/N   _____

    Knives   Y/N   _____

    Slicers
                        Y/N   _____

    Work surfaces
                        Y/N   _____

    Prep sinks

    Pocket thermometers   Y/N   _____

    Thermometers sanitized each use Y/N   _____

Describe system and frequency of thermometer calibration.

_____

_____

_____

_____

_____

_____

COUNTY	DIST.	EST. NO.	MONTH	DAY	YEAR

THIS FORM CONSISTS OF TWO PAGES AND BOTH MUST BE COMPLETED.

Establishment Name _____   Operator's Name _____

Address _____

(T) (C) (V) _____   County _____

Food _____

PROCESS (STEP) CIRCLE CCPs	CRITERIA FOR CONTROL	MONITORING PROCEDURE OR WHAT TO LOOK FOR	ACTIONS TO TAKE WHEN CRITERIA NOT MET
RECEIVING/ STORAGE	☐ Approved source (inspected) ☐ Shellfish tag ☐ Raw/Cooked/Separated in storage ☐ Refrigerate at less than or equal to 40°F	☐ Shellfish tags available ☐ Shellfish tags complete ☐ Measure food temperature ☐ No raw foods stored above cooked or ready-to-eat foods	☐ Discard food ☐ Return food ☐ Separate raw from cooked food ☐ Discard cooked food contaminated by raw food ☐ Food Temperature More than 40°F more than 2 hours, discard food ☐ More than 70° F, discard food
THAWING	☐ Under refrigeration ☐ Under running water less than 70°F ☐ Microwave ☐ Less than 3 lbs., cooked frozen ☐ More than 3 lbs., do not cook until thawed	Observe method Measure food temperature	Food temperature: More than or equal to 70° F, discard More than 40°F more than 2 hours, discard
PROCESSING PRIOR TO COOKING	Food temperature less than or equal to 45°F	Observe quantity of food at room temperature Observe time food held at room temperature	Food temperature: More than 40°F more than 2 hours, discard food More than 70° F, discard food
COOKING	Temperature to kill pathogens Food temperature at thickest part more than or equal to _____°F	Measure food temperature at thickest part	Continue cooking until food temperature at thickest part is more than or equal to _____°F
HOT HOLDING	Food temperature at thickest part more than or equal to _____°F	Measure food temperature at thickest part during hot holdng every _____ minutes	Food temperature: 140°F–120°F More than or equal to 2 hours, discard; less than 2 hours, reheat to 165°F and hold at 140°F 120°F–40°F More than or equal to 2 hours, discard; less than 2 hours, reheat to 165°F and hold at 140°F

Food_____ Establishment Name_____ Date_____

PROCESS (STEP) CIRCLE CCPs	CRITERIA FOR CONTROL	MONITORING PROCEDURE OR WHAT TO LOOK FOR	ACTIONS TO TAKE WHEN CRITERIA NOT MET
COOLING	Food 120°F to 70°F in hours: 70°F to 40°F in 4 additional hours by the following methods: (check all that apply) ☐ Product depth less than or equal to 3 inches ☐ Ice water bath and stirring ☐ Solid pieces less than or equal to 6 lbs. ☐ Rapid chill all ingredients ☐ No covers until cold	Measure temperature during cooling every _____ minutes ☐ Food depth ☐ Food iced ☐ Food stirred ☐ Food size ☐ Food placed in rapid chill refrigeration unit ☐ Food uncovered	Food temperature: 120°F–70°F more than 2 hours, discard food 70°F–40°F more than 4 hours, discard food 40°F or less but cooled too slowly, discard food
PROCESSING SLICING DEBONING MIXING DICING ASSEMBLING SERVING	Prevent contamination by: Ill workers not working Workers hands not touching ready-to-eat foods Workers hands washed Cold potentially hazardous food at temperatures less than or equal to 40°F Hot potentially hazardous food at temperature more than or equal to 140°F Equipment and utensils clean and sanitized	Observe: Workers' health Use of gloves, utensils Handwashing technique Wash & sanitize equipment & utensils Use prechilled ingredients for cold foods Minimize quantity of food at room temperature Measure food temperature	If yes to following, discard: Ill worker working Direct hand contact with ready-to-eat food observed Cold potentially hazardous food: more than 40°F more than or equal to 2 hours, discard; more than 70°F, discard Hot potentially hazardous food 140°F–120°F More than or equal to 2 hours, discard; less than 2 hours, reheat to 165°F and hold at 140°F If yes to following, discard or reheat to 165°F: Raw food contaminated by other food Equipment/utensils are contaminated
REHEATING	Food temperature at thickest part more than or equal to 165°F	Measure food temperature during reheating	Food temperature less than 165°F, continue reheating
HOLDING FOOD, HOT/COLD TRANSPORTING FOOD	Food temperature ☐ More than or equal to 140°F at thickest part ☐ Less than or equal to 40°F at thickest part	Measure food temperature during holding every _____ minutes	☐ Hot holding potentially hazardous food: 140°F–120°F More than or equal to 2 hours, discard; less than 2 hours, reheat to 165°F and hold at 140°F 120°F–40°F More than or equal to 2 hours, discard; less than 2 hours, reheat to 165°F and hold at 140°F ☐ Cold holding potentially hazardous food temperature: 40°F–70°F More than or equal to 2 hours, discard; less than 2 hours, serve or refrigerate More than or equal to 70°F, discard

I have read the above food preparation procedures and agree to follow and monitor the critical control points and to take appropriate corrective action when needed. If I want to make any changes, I will notify the Health Department prior to such a change.

Signature of person in charge_____ Signature of inspector_____

STEP 7

Establish Procedures for Verification

Conduct periodic audits to make sure the HACCP system works.

Conduct an audit of the entire HACCP system at least annually.

Conduct additional audits whenever there are new products, new recipes, or a new process. Each of these requires a new HACCP plan.

Surrender
is an operation
by means of
which we set
about explaining
instead of
acting.

–Charles Pequy

APPLICATION OF STEP 7

Establish Procedures for Verification

Verification procedures may include:

Establishment of appropriate inspection schedules

Review of the HACCP plan

Review of CCP records

Review of deviations and corrections

Visual inspection of operation to observe whether CCPs are under control

Random sample collection and analysis

Review of critical limits to verify that they are adequate to control hazards

Review of written record of verification inspections covering compliance, deviations, and corrective action taken

Review of modification of the HACCP plan

Components of an Effective HACCP Plan*

The following lists describe the components essential to an effective HACCP plan.

Management, Supervisor, and Employee Responsibility for Food Safety

1. Demonstrate *management commitment* through food safety promotions: safety committee, incentives, awards.

     Set food safety goals that are challenging, measurable, and attainable.
     Demonstrate food safety examples in all activities.
     Discuss and respond to employee suggestions for improvements.

2. Develop an *organizational chart* showing assigned responsibilities for specific workplace food hazard controls.

3. Develop *food safety committees*.

4. Establish *accountability* for meeting food safety responsibilities.

5. Hold regular *staff meetings* to reinforce the safety principles and listen for employee suggestions for improvement.

6. Conduct *mini "how to" demonstrations* for hand washing, cool-down procedures, sanitizing solutions, and calibration of thermometers.

7. Implement *ongoing inspection and monitoring* programs to identify and improve controls of changing workplace hazards.

---

*HACCP-Based Total Quality Management Hospital Foodservice, Peter Snyder, Jr., Hospitality Institute of Technology and Management.

Hazard Analysis and Control

1. Identify and analyze *workplace food hazards* that could lead to foodborne illness through food safety self-inspections.

2. Examine *each food safety hazard* and complete the following:

   Flow chart the process, identifying ingredients, time, temperature, and equipment essential to hazard control.

   Identify critical control points at which there are potential food safety hazards that can lead to illness.

   Establish procedures that, when used, will control hazards.

3. Schedule an available *hazard control quality-assured manager* on all shifts.

4. Maintain effective *record-keeping systems.*

Written Food Safety Goals and Objectives

1. Write a food safety *policy statement* to control hazards specific to the workplace.

2. Write a *food safety action plan and program,* clearly describing how food safety assurance (precontrol), safety control, and safety improvement goals will be met.

3. Write plans for conducting and documenting a *review* of the program effectiveness, and for improving the program based on findings.

4. *Purchase* equipment that meets the need for food safety; for example, shallow pans for cooling, thermometers for refrigeration, readily available probe thermometers.

5. Institute appropriate *equipment programs* to cover the calibration, use, cleaning, maintenance, and eventual replacement of equipment.

Communication and Training

1. Communicate the *food safety program* to all employees.

2. Allow for *employee input* in bringing hazardous food operating conditions to management's attention.

3. Provide *training prior to all new job assignments,* including training on specified hazard controls.

4. *Update training* at least annually or as work procedures and ingredients change.

5. Maintain *records of training:* date, topic, content, attendance, and trainer.

6. *Train supervisors* in pertinent food safety matters, food safety leadership, coaching, and employee training to bring food safety problems to the attention of the supervisor.

7. Evaluate *training needs* to determine specialized training and retraining. Encourage supervisors and employees to give feedback as to how to improve training.

Establish Investigation and Corrective Action Procedures

1. Develop *procedures* for problem reporting, problem investigation, corrective action, and follow-up.

2. Conduct *workplace prevention* inspection of facilities and equipment (e.g., refrigeration, cooking, and hot holding devices; pot and dish washing and sanitizing; insect and rodent control).

3. Write reports following the control of problems, showing what *preventive/corrective action* is being taken to prevent similar problems; for example:
   Equipment modified
   Work method modified
   Change recipe
   Different ingredient
   Equipment changed or added
   Employee retrained or special needs accommodated

4. Maintain, summarize, and *analyze foodborne illness data* (e.g., first reports of illness) to determine tasks and operations in which incidents have occurred. Take corrective action to prevent recurrence.

Program Enforcement

1. Write an *enforcement statement* on safe food operation practices, food safety rules, and standard operating procedures.

2. Maintain *records of disciplinary action and warnings.*

3. Develop policies that hold *managers, supervisors, and employees accountable* for food safety responsibilities. Hold everyone responsible for his or her behavior in adhering to safety standards.

Management Self–Inspection/Verification Procedures

To verify that the HACCP plan is working, review your HACCP system at least once a year. The following is an outline procedure for conducting an audit of your HACCP system. This format is also a useful tool for conducting "hands on HACCP training" for the management team or for a classroom field experience.

1. Review the menus with the manager/cook and ask him or her to describe the preparation steps. While reviewing the menu, use some of the questions from pages 214–215 to determine potential hazards.

2. Focus on those foods that meet the definition for potentially hazardous food (PHF), note critical control points (CCPs), and construct a trial flow chart of one or two food products.

3. Determine when preparation of PHF begins and schedule your inspection to start as close to that time as possible.

4. Locate and inspect the ingredients to be used in PHF.

5. Locate and inspect any leftover PHF product on hand that will not be used up before the HACCP inspection. The refrigerated storage is the best place to start this step. Record temperatures of the product on hand.

HACCP Inspection

6. Start your HACCP inspection as scheduled in Step 3. Observe and learn the preparation steps. Use some of the questions from pages 214–215 to develop an understanding of the current handling processes and procedures. Use the HACCP Monitoring Procedure Report form on pages 216–217 as part of the audit process. This form focuses on the critical items in the HACCP system.

- Record product temperatures at *various* stages of preparation. Monitor product temperature during storage. (1–2 hour intervals).

- Be alert for potential cross-contamination from hands, raw products, surfaces, wiping cloths, and so forth.

- Observe the process and track food and containers through the process.

Inspection Tips

Review the menu, recipes, and policy guides.

Ask questions. Be sure you understand the complete operation.

Calibrate your equipment.

Follow the food through the handling process. Develop accurate flow charts.

Verify that the food in question meets the legal definition for potentially hazardous food.

Document times and temperatures.

Be aware of possibilities for cross-contamination (hands, towels, gloves, cutting boards, slicers, utensils).

Avoid making recommendations and corrections during the inspection. Save your comments for your presentation.

- Record all product time/temperatures on the work sheet provided. See the sample work sheet on page 219.
- Sketch a rough kitchen layout and record the flow of various raw foods, cooked foods, and soiled articles throughout the kitchen.
- Take pictures (optional) to document procedures or practices that may be hazards or critical control points (CCPs).
- Document specific violations. Example: Raw chicken juice dripping into prepared food.

7a. Prepare a flow chart to reflect *actual* preparation steps. A sample HACCP flow chart is provided on page 213. Based on the *data you have collected*, identify the potential hazards and critical control points on each PHF flow chart. A critical control point is a point where a hazard can be prevented or controlled and that provides an opportunity for monitoring.

Prepare a Time/Temperature graph from the data collected on the same Time/Temperature Survey Data Form. Samples of these forms are on pages 218–220.

b. When conducting an audit in a facility that has an established HACCP system, also review:
- The written HACCP plan
- CCP records
- Records of deviations and corrections
- Written records of verification inspections
- Employee training records

Group Presentations

If you are using this format for classroom or management training, continue with Step 8.

8. Transfer your flow chart to a flip chart or overhead sheet and present your findings to the rest of the group. Keep critical comments short and on the subject of CCP. Be sure to give positive comments as well.

Final Report

9. If you were guests in a facility that allowed you to use their operation as a field experience, as a consideration provide them with a written report of your findings and a thank-you letter. The report should consist of:
- Final flow charts with hazards and critical control points noted on them.
- Time-temperature work sheet or graph of data from work sheet (see sample on pages 218, 220).
- Recommendations
- Other observations (optional)

10. Send copies of the final report of your observations to the manager of the host facility.

HACCP FLOW CHART
(Mishandled Product)

Establishment Name _____ ABC Restaurant _____

Time/Date	Procedure	Comments
4/1/94	Frozen Chicken	
9:00 p.m.	Thaw in Refrigerator	
	Bread	
9:30 p.m. **CCP***	Cook in Oil	Internal Product Temp. Reached 180°F.
9:45 p.m. **CCP***	Hold Hot in Warmer	Internal Product Temp. 112°F. Product Time Coded for Discard after 2 Hrs.
	Sell	
10:00 p.m. **CCP***	Cool Overnight	Leftover Product Placed in 6-inch Deep Pans, Covered with Foil, Stacked 3 Deep in Refrigerator. Recording Thermometer Verified It Took 7 Hrs. to Cool to 40°F.
4/2/94 9:00 a.m. **CCP***	Leftover Reheated	Reheat Temp. Reached Only 112°F in Food Warmer.
9:30 a.m.	Sell	

Critical control point in violation of food safety concepts.

Questions Regulators Ask

When a health inspector from a governmental regulatory agency visits a facility, he or she needs to learn about the operation. In the menu review process with the manager or chef, the regulator will ask questions to identify potential hazards. The next step is to observe the food flow. While observing, the regulator will ask more questions of food handlers. The answers will provide better understanding of the practices and procedures and complete the food flow particulars. The following are typical questions asked by regulators visiting food-service operations:

Source

1. Do you cater outside the establishment?
2. Are any foods that are served here prepared at anyone's home?

Food Protection

3. Do you use raw eggs in any recipes? What final cooking temperature is specified?
4. What recipes call for ground meat? What final cooking temperature is specified?
5. What dishes use fresh raw chicken or turkey? When, where, preparation by whom, etc.?
6. Do you serve raw seafood? What, how, from where, prepared by whom? What about storage/transportation temperatures, rotation? Are you aware of recommendations for immune-compromised food handlers? Are you aware of the symptoms of food-borne illness? Do you have training materials? Would you be interested in obtaining some?
7. What is your policy on employee illness? Have employees received training on illness symptoms and disease transmission? Are special precautions taken for personal hygiene?
8. What is your policy on cuts and burns?
9. Do you check the accuracy of your food thermometers? How?
10. Do you serve lightly cooked egg products, such as scrambled eggs, omelets, and French toast? If you are using fresh eggs, are you aware of the salmonella e. problem?
11. What foods are cooked a day or more in advance of service? How are they cooled? How are they reheated? Have you actually monitored these foods?
12. Do you cook and bone chickens or turkeys?
13. Do you have a policy/procedure for addressing complaints of injury or illness? Do you have a designated person to receive such complaints? Have you received any complaints in the last six months? Do you want any assistance in figuring out what happened?

14. Where, when, how, and by whom are the following items cleaned and sanitized?
    Chef's knives
    Bakery department tools
    Salad department tools
    Other general items—pots, pans, etc.

15. What are your instructions to employees about wiping cloths? What is your system to assure the correct sanitizer is being used in the appropriate concentration and that the solutions are being monitored and changed?

16. Do you cook ground meat patties that are frozen or thick (such as half-pound patties) or meat loaf? What final cooking temperature is specified? Are you using any new techniques like sous vide, cook/chill, or vacuum packaging? Are you aware of the new cooking recommendations for hamburger patties?

17. What recipes incorporate leftovers (i.e., fried chicken, baked potato, baron of beef)? What final reheating temperatures are specified?

18. Where are salad ingredients washed? Are the sinks where vegetables are washed also used for raw meats, poultry, or hand washing? If so, what are the special instructions or precautions?

19. Do you have any cleaning or sanitizing products that resemble food products as to label, color, or odor?

20. If pests, such as cockroaches, are or have been present, how are they controlled? By operator? By a pest control operator (PCO)? Using what compounds? How frequently? By what method? Do you have receipts? Have you accompanied the PCO? Do you know what the law requires? Have you read the labels for the products being used in your establishment?

21. Do you serve precooked meat patties or hot dogs? What final temperature are specified? Are you aware of the new cooking requirements?

22. To what temperature do you cook pork? Are you aware of the new cooking recommendations?

TEMPERATURE ANSWERS:

3.	145°F
4.	165°F
16.	155°F
17	165°F
21.	140°F
22.	155°F

Hazard Analysis Critical Control Point
Monitoring Procedure Report

NEW YORK STATE DEPARTMENT OF HEALTH
Bureau of Community Sanitation and Food Protection

HACCP Page 1

COUNTY	DIST.	EST. NO.	MONTH	DAY	YEAR

THIS FORM CONSISTS OF TWO PAGES AND BOTH MUST BE COMPLETED.

Establishment Name _____ Operator's Name _____

Address _____

(T) (C) (V) _____ County _____

Food _____

PROCESS (STEP) CIRCLE CCPs	CRITERIA FOR CONTROL	MONITORING PROCEDURE OR WHAT TO LOOK FOR	ACTIONS TO TAKE WHEN CRITERIA NOT MET
RECEIVING/ STORAGE	☐ Approved source (inspected) ☐ Shellfish tag ☐ Raw/Cooked/Separated in storage ☐ Refrigerate at less than or equal to 40°F	☐ Shellfish tags available ☐ Shellfish tags complete ☐ Measure food temperature ☐ No raw foods stored above cooked or ready-to-eat foods	☐ Discard food ☐ Return food ☐ Separate raw from cooked food ☐ Discard cooked food contaminated by raw food ☐ Food Temperature More than 40°F more than 2 hours, discard food ☐ More than 70° F, discard food
THAWING	☐ Under refrigeration ☐ Under running water less than 70°F ☐ Microwave ☐ Less than 3 lbs., cooked frozen ☐ More than 3 lbs., do not cook until thawed	Observe method Measure food temperature	Food temperature: More than or equal to 70° F, discard More than 40°F more than 2 hours, discard
PROCESSING PRIOR TO COOKING	Food temperature less than or equal to 45°F	Observe quantity of food at room temperature Observe time food held at room temperature	Food temperature: More than 40°F more than 2 hours, discard food More than 70° F, discard food
COOKING	Temperature to kill pathogens Food temperature at thickest part more than or equal to _____°F	Measure food temperature at thickest part	Continue cooking until food temperature at thickest part is more than or equal to _____°F
HOT HOLDING	Food temperature at thickest part more than or equal to _____°F	Measure food temperature at thickest part during hot holdng every _____ minutes	Food temperature: 140°F–120°F More than or equal to 2 hours, discard; less than 2 hours, reheat to 165°F and hold at 140°F 120°F–40°F More than or equal to 2 hours, discard; less than 2 hours, reheat to 165°F and hold at 140°F

216

Food_____ Establishment Name_____ Date_____

PROCESS (STEP) CIRCLE CCPs	CRITERIA FOR CONTROL	MONITORING PROCEDURE OR WHAT TO LOOK FOR	ACTIONS TO TAKE WHEN CRITERIA NOT MET
COOLING	Food 120°F to 70°F in hours: 70°F to 40°F in 4 additional hours by the following methods: (check all that apply) ☐ Product depth less than or equal to 3 inches ☐ Ice water bath and stirring ☐ Solid pieces less than or equal to 6 lbs. ☐ Rapid chill all ingredients ☐ No covers until cold	Measure temperature during cooling every _____ minutes ☐ Food depth ☐ Food iced ☐ Food stirred ☐ Food size ☐ Food placed in rapid chill refrigeration unit ☐ Food uncovered	Food temperature: 120°F–70°F more than 2 hours, discard food 70°F–40°F more than 4 hours, discard food 40°F or less but cooled too slowly, discard food
PROCESSING SLICING DEBONING MIXING DICING ASSEMBLING SERVING	Prevent contamination by: Ill workers not working Workers hands not touching ready-to-eat foods Workers hands washed Cold potentially hazardous food at temperatures less than or equal to 40°F Hot potentially hazardous food at temperature more than or equal to 140°F Equipment and utensils clean and sanitized	Observe: Workers' health Use of gloves, utensils Handwashing technique Wash & sanitize equipment & utensils Use prechilled ingredients for cold foods Minimize quantity of food at room temperature Measure food temperature	If yes to following, discard: Ill worker working Direct hand contact with ready-to-eat food observed Cold potentially hazardous food: more than 40°F more than or equal to 2 hours, discard; more than 70°F, discard Hot potentially hazardous food 140°F–120°F More than or equal to 2 hours, discard; less than 2 hours, reheat to 165°F and hold at 140°F If yes to following, discard or reheat to 165°F: Raw food contaminated by other food Equipment/utensils are contaminated
REHEATING	Food temperature at thickest part more than or equal to 165°F	Measure food temperature during reheating	Food temperature less than 165°F, continue reheating
HOLDING FOOD, HOT/COLD TRANSPORTING FOOD	Food temperature ☐ More than or equal to 140°F at thickest part ☐ Less than or equal to 40°F at thickest part	Measure food temperature during holding every _____ minutes	☐ Hot holding potentially hazardous food: 140°F–120°F More than or equal to 2 hours, discard; less than 2 hours, reheat to 165°F and hold at 140°F 120°F–40°F More than or equal to 2 hours, discard; less than 2 hours, reheat to 165°F and hold at 140°F ☐ Cold holding potentially hazardous food temperature: 40°F–70°F More than or equal to 2 hours, discard; less than 2 hours, serve or refrigerate More than or equal to 70°F, discard

I have read the above food preparation procedures and agree to follow and monitor the critical control points and to take appropriate corrective action when needed. If I want to make any changes, I will notify the Health Department prior to such a change.

Signature of person in charge _____ Signature of inspector _____

SAMPLE WORK SHEET
Time/Temperature Survey Data

Date: _____

Product: _____ Height: _____ Length: _____

Storage Station: _____ Width: _____ Cr. Diameter: _____

Ambient Temperature: _____ of Storage Pot, Pan, or Container

Time	Temperature	Time	Temperature
_____	_____	_____	_____
_____	_____	_____	_____
_____	_____	_____	_____
_____	_____	_____	_____
_____	_____	_____	_____
_____	_____	_____	_____
_____	_____	_____	_____
_____	_____	_____	_____
_____	_____	_____	_____
_____	_____	_____	_____
_____	_____	_____	_____
_____	_____	_____	_____
_____	_____	_____	_____

FLOW CHART

Product: _____

Date: _____

Time/Temperature Graph

Date: _____

Product: _____

Container: _____

Method of Cooling: _____

Temp:

170°	
165°	
160°	
155°	
150°	
145°	
140°	
135°	
130°	
125°	
120°	
115°	
110°	
105°	
100°	
95°	
90°	
85°	
80°	
75°	
70°	
65°	
60°	
55°	
50°	
45°	
40°	

Time (hours)

220

Foodborne Illness

Results of Foodborne Illness

- Loss of Customers and Sales

- Loss of Prestige and Reputation

- Legal Suits Resulting in Lawyer and Court Fees

- Increased Insurance Premiums

- Lowered Employee Morale

- Absenteeism of Employees

- Need for Retraining Employees

- Embarrassment

*H*uman history
becomes
more and more
a race
between education
and
catastrophe.

–*H.G. Wells*

FOODBORNE ILLNESS

40 TO 81 MILLION
CASES PER YEAR

10,000 DEATHS PER YEAR

Economic Effects of Foodborne Illness

It is estimated that up to 81 million cases of foodborne illness occur each year resulting in 10,000 deaths. The costs incurred by the foodservice industry as a result of these outbreaks include, but are not limited to: medical care, lost wages, public health investigation, lost business, and legal action.

An investigator who studied 17 foodborne illness outbreaks in the United States, Canada, and other countries found an average cost of almost $200,000 per outbreak.

Others have estimated that diarrheal foodborne illness costs between $7.7 and $23 billion a year in the United States. These figures show that everyone has an economic interest in preventing foodborne illness.*

FIVE FACTORS THAT CAUSE 80% OF FOODBORNE ILLNESS OUTBREAKS

Improper Cooling	30%
Advance Preparation	17%
Infected Person	13%
Inadequate Reheating	11%
Improper Hot Holding	9%

The Pareto Diagram on the following pages graphically illustrates the factors that contribute to outbreaks of foodborne illness as the result of mishandling of foods in foodservice establishments.

---

*Bacteria That Cause Foodborne Illness, U.S. Department of Agriculture, December 1990. Food Code 1993, U.S. Public Health Service, FDA.

PARETO DIAGRAM

Factors Contributing to Outbreaks of Foodborne Illness in Foodservice Establishments

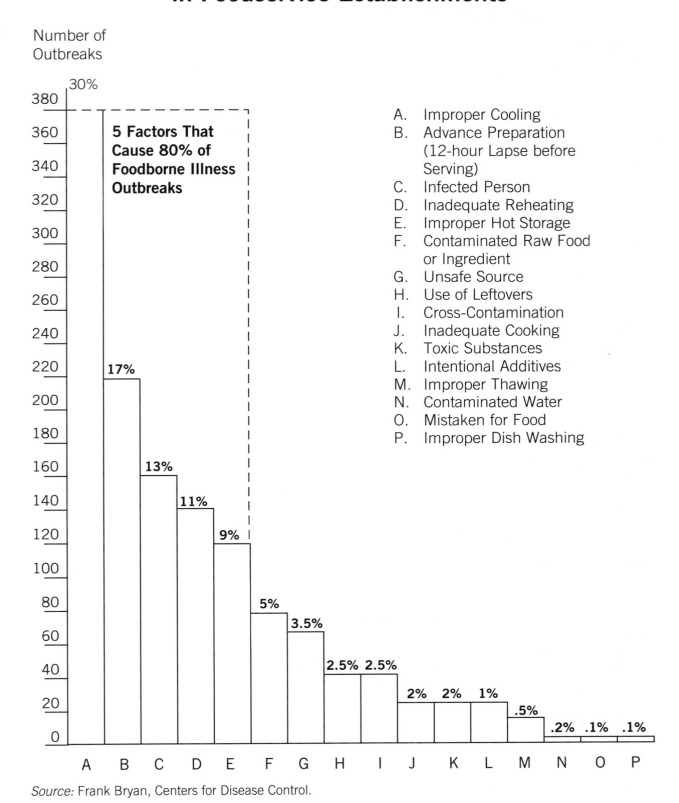

Number of Outbreaks

5 Factors That Cause 80% of Foodborne Illness Outbreaks

A. Improper Cooling
B. Advance Preparation (12-hour Lapse before Serving)
C. Infected Person
D. Inadequate Reheating
E. Improper Hot Storage
F. Contaminated Raw Food or Ingredient
G. Unsafe Source
H. Use of Leftovers
I. Cross-Contamination
J. Inadequate Cooking
K. Toxic Substances
L. Intentional Additives
M. Improper Thawing
N. Contaminated Water
O. Mistaken for Food
P. Improper Dish Washing

Source: Frank Bryan, Centers for Disease Control.

PARETO DIAGRAMA

Factores que Contribuyen a Enfermedades Causados del Malmanejo y/o maltratamiento de Comida Erupciones en Establecimientos de Alimientos

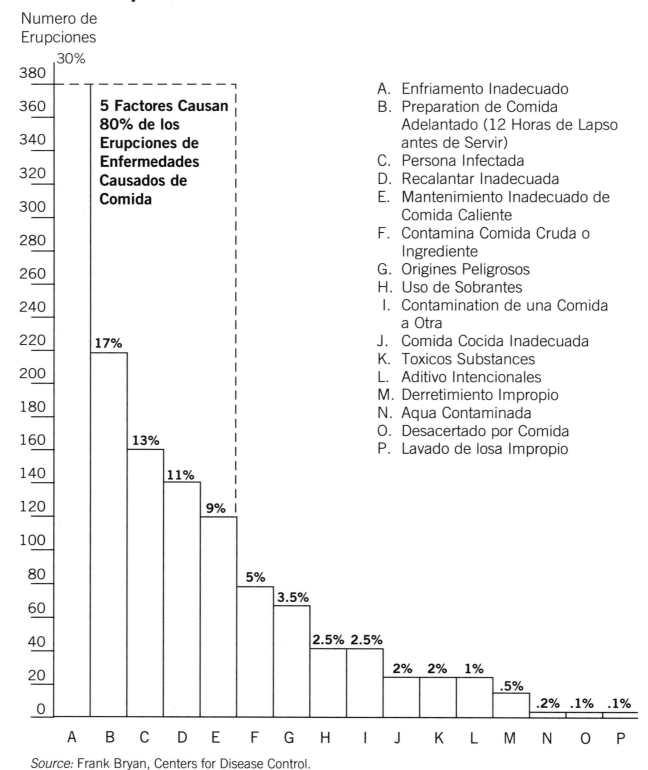

A. Enfriamento Inadecuado
B. Preparation de Comida Adelantado (12 Horas de Lapso antes de Servir)
C. Persona Infectada
D. Recalantar Inadecuada
E. Mantenimiento Inadecuado de Comida Caliente
F. Contamina Comida Cruda o Ingrediente
G. Origines Peligrosos
H. Uso de Sobrantes
I. Contamination de una Comida a Otra
J. Comida Cocida Inadecuada
K. Toxicos Substances
L. Aditivo Intencionales
M. Derretimiento Impropio
N. Aqua Contaminada
O. Desacertado por Comida
P. Lavado de losa Impropio

Source: Frank Bryan, Centers for Disease Control.

What Causes Foodborne Illness?

As shown in the table on page 253, there are at least 12 causes of foodborne illness. The diagram on page 225 illustrates that five factors cause 80 percent of the cases. Four of the five factors relate to time/temperature. These four time/temperature factors are:

- Improper cooling
- Preparation 12 hours or more in advance of serving
- Inadequate reheating
- Improper hot storage

The fifth critical factor relates to food handling by infected persons. This accounts for 13 to 30 percent of foodborne illness. (The exact percentage varies depending on the study and how the information is analyzed.)

If these five factors cause 80 percent of foodborne illness, then a foodservice operator can significantly reduce the risk by concentrating on three areas:

- Employee health and hygiene practices
- Proper temperature control and monitoring during cooking, cooling, holding, and reheating
- Proper use and concentration of sanitizers

By applying the HACCP steps and understanding basic microbiology, foodservice managers can set up practical procedures and effective training of food handlers to minimize the risk of foodborne illness in their establishments.

Factors Contributing to Outbreaks of Foodborne Illness

1. Improper cooling

2. Lapse of 12 hours or more between preparation and service of food

3. Employees who are infected or who practice poor personal hygiene

4. Inadequate cooking

5. Improper hot storage

6. Inadequate reheating: Use of leftovers

7. Contaminated raw foods

8. Cross-contamination

9. Improper cleaning and sanitizing of equipment

10. Toxic substances

11. Unsafe food sources

12. Poor food storage practices

Factors Contribuyen tes Enfermedades Causadas por Comidas Contaminadas

1. Enfriamento inadecuado

2. El transcurso de 12 horas o mas entre la preparacion y el servicio de la comida

3. Empleados infectados o los que practican una higiene personal inadecuada

4. Cocinar los alimentos inadecuadamente

5. El mantenimento impropio de la comida caliente

6. Inadecuado recalentamiento incorrecto o el uso de sobrantes de comida

7. Comida cruda contaminada

8. Contaminacion de una comida a otra

9. Limpieza inadecuada y desinfeccion inadecuado del equipo

10. Substancias toxicas

11. Origenes no autarizados de los alimentos

12. Almacenamiento incorrecto de los alimentos

Factors Necessary for an Outbreak to Occur

Foodborne illness is an illness caused by eating food that contains an *infective dose* of disease-causing organisms. It can strike if three things are present:

1. Food is contaminated as a result of:
    Infected persons
    Poor personal hygiene
    Contaminated raw food or ingredient
    Unsafe source
    Cross-contamination

2. The contaminated food remains in a favorable state for growth of the bacteria or virus as a result of:
    Improper cooling
    Advance preparation
    Inadequate reheating
    Improper hot holding
    Improper thawing

3. An individual consumes an *infective dose* of the contaminated food. An infective dose is the number of organisms needed to make a person ill from a particular bacteria or virus. According to the Denver Department of Public Health and Consumer Protection, infective doses of some key organisms are:

Organisms	Dose
Shigella/E.coli	As few as 10 cells
C. Perfringens/Staphylococcus	100,000/g
Salmonella	100–1000
Campylobacter	>500

Sometimes only one individual complains of food poisoning. If an infective dose is present, why is there only one complainant? This is possible because some people are immune compromised. Their immune systems are less effective because, for example, they are pregnant, elderly, or very young, have recently been ill, or are taking medications. Also, bacteria grow where they are most comfortable and different people provide different hosts. And then, of course, people don't always report it when they become ill.

How Do Organisms Get into Food?

Harmful organisms are introduced into food in three main ways:

As a Result of Inadequate Processing
Samonella/Campylobacteria in poultry

E. coli in ground beef

Listeria in unpasteurized milk

From Person to Person
Shigella/Hepatitis A (fecal-oral route)

Staphylococcus (common on skin, throat, nose)

From Food to Food
Handling raw, then ready-to-eat food

Use of common cutting boards

Improper storage of potentially hazardous foods

The Biology of Foodborne Illness

An understanding of basic microbiology is important to hazard analysis. To properly assess the points in a food preparation flow, you must be able to make some decisions about product contamination and, more importantly, the potential for bacterial growth.

Not all food hazards are bacterial in nature. However, reports to the Centers for Disease Control indicate that bacterial agents are the leading cause of foodborne illness in the United States today. The table on page 247 lists five main bacteria and one virus that cause illness, with their sources, the resulting diseases, and control measures.

> *N*ewton and Murphy were right: Things left to themselves generally go from bad to worse.
>
> *–Author Unknown*

Infections

Some bacteria make people sick when they eat live harmful organisms. This is called an *infection*. Examples of bacterial infections are salmonellosis, shigellosis, and camplobacteriosis. These bacteria are killed by cooking a product to its designated cooking temperature. For example, most salmonella bacteria are killed when food is cooked to 165°F. Eighty-eight percent of foodborne illness is caused by bacteria.

Intoxication

Other bacteria make people ill from their waste products. This type of foodborne illness is called *intoxication*. Examples of intoxication-causing bacteria are *Bacillus cereus, Staphylococcus aureus, Clostridium perfringens,* and *Escherichia coli.* The toxins from *B. cereus* and *staph* are not destroyed by heat. Heating food products containing these bacterial toxins will not prevent foodborne illness.

B. cereus and *C. perfringens* form spores that can survive freezing and boiling temperatures. These bacteria are controlled by rapid cooling and reheating of food to prevent their growth in food and prevent time for spore formation. Avoiding contamination is also required for control.

The toxin of *E. coli* is not heat stable, so cooking food to 155°F, maintaining cold storage temperatures, practicing good personal hygiene standards, and prevention of cross-contamination will control the bacteria.

Viruses

Viral illness is transmitted by poor personal hygiene. Food, water, and air can become the vehicles by which the disease is spread. Viruses, however, do not grow in food. The Norwalk and hepatitis A viruses are spread by fecal contamination of infected workers. Proper hand washing is required to control the spread of these viral infections. Five percent of foodborne illnesses are caused by viruses.

Infected Workers

Infected food handlers may have mild illnesses, or no symptoms at all, and still shed salmonella or hepatitis A in their stools. That is why personal hygiene is such an important control for these and other organisms. Any employee with diarrhea should not be allowed to handle foods or clean utensils, and you may wish to request medical clearance before an ill employee returns to work.

Bacteria-Caused Foodborne Illness*

INTOXICATION	INFECTION
Staphylococcus (infected food handler)	Salmonella (chicken)
Bacillus (cooked rice)	Shigella (fecal contamination)
C. Perfringens (produce, soil)	Campylobacter (chicken)
E. coli (ground beef)	
Shorter incubation period: 1–12 hours	Longer incubation period: 6–72 hours

Denver Department of Public Health/Consumer Protection.

Growth Patterns

Bacteria require certain conditions in order to multiply rapidly. It is rapid bacteria multiplication that often causes problems with regard to the safety of a food product. Under ideal conditions, rapid growth can mean that an organism has a generation time of as little as 20 minutes. The following example assumes that a certain food initially contains 1,000 organisms per gram.

Time	Number of Organisms
1 Hour later	4,000
2 Hours	16,000
3 Hours	64,000
4 Hours	256,000
5 Hours	1,024,000

This ideal rapid growth is called the *log phase,* and all bacteria will reach this rapid part of their growth if given the correct conditions. Bacteria begin the growth cycle by adjusting to any new environment or conditions in a *resting* or *lag phase.* It is at the lag phase that bacterial growth can be best controlled for preventing foodborne illness. *Stationary* and *death phases* are usually brought about by the depletion of available nutrients and the production of metabolic waste products (intoxication). Bacterial growth cycles look like the following graph below.

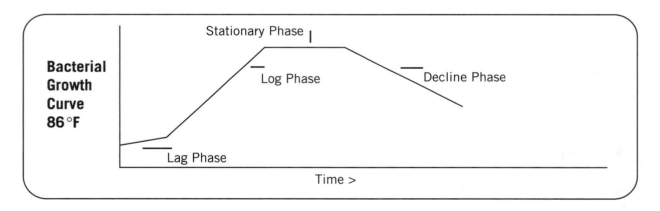

Factors Affecting Growth

The length of the lag phase and the slope of the log phase are affected by the following environmental factors:

Temperature
Nutrients
Water activity
pH (alkalinity–acidity)
Atmosphere
Presence of inhibiting substances
Time

TEMPERATURE

Temperature is one of the factors that can be used to "manage" the number of organisms that may be present in a food product. Temperature is a well-known management tool to keep bacteria levels low.

The following graph shows the change in the predicted growth curves for salmonella bacteria at 60°F versus the first growth curve under ideal conditions at 86°F (see graph under "Growth Patterns" on page 234). The lag phase at the lower temperature has been considerably lengthened.

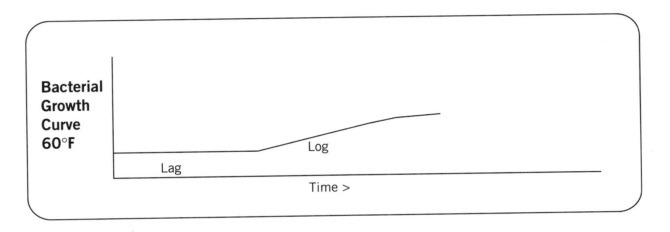

Similar "stretching" of the lag phase occurs as other environmental parameters are moved away from ideal environment. The pH, water activity, and other factors can be used to keep bacteria from multiplying rapidly in food.

pH

The pH of a food product is the measure of its acidity or alkalinity. The pH scale begins at zero and ends at 14. A solution with a pH of 7.0 is considered neutral, neither acidic nor alkaline. Pure distilled water would show a pH of 7.0.

Microorganisms of concern to food safety like to grow in surrounding that have a pH near this neutral measure. As a product's pH moves below or above the ideal level for a particular organism, the microbe takes longer to adjust to its surroundings (lag phase of growth) and grows more slowly.

Many foods are naturally acidic, such as vinegar, mayonnaise, fruits, pickles, and yogurt, and have a pH below 4.6. At or below 4.6, disease-causing organisms do not grow or grow so slowly that they are not a food safety problem. Spoilage organisms, however, may grow at these low pH values and can slowly change a food's taste or appearance.

The pH of a food has a great deal of effect on its suitability as a medium for bacterial growth. Knowing the pH of a food item can aid in determining the critical controls needed to maintain its safety.

The following tables show the approximate range of pH values at which organisms grow and the ranges of some foods. These illustrations can give you an idea as to how you can include this parameter of bacterial growth in your HACCP tool bag.

MICROORGANISMS AND pH

Organisms	Optimum pH
Salmonella	6.0–7.5
Staphylococcus	6.8–7.5
E. Coli	6.0–8.0
Most Bacteria	5.5–8.0
Yeast (spoilage organisms)	4.0–6.5
Molds (spoilage organisms)	4.5–6.8

pH VALUES OF SOME FOODS

Product	pH
Ground Beef	5.1–6.3
Ham	5.9–6.1
Chicken	6.2–6.7
Fish	6.6–6.8
Clams	6.5
Crabs	7.0
Oysters	4.8–6.3
Butter	6.1–6.4
Buttermilk	4.5
Cheese	4.9–5.9
Milk	6.6–7.0
Yogurt	3.8–4.2
Vegetables	4.2–6.5
Onions	4.8
Tomatoes, Fresh	4.2–4.9
Tomatoes, Canned & Paste	3.5–4.7
Fruits	2.0–6.7
Orange Juice	4.0
Grapefruit	3.6
Mayonnaise	3.0–4.1
Salad Dressing	3.2–4.0

Note: Values will vary by source.

WATER ACTIVITY (A$_w$)

Bacteria need water in an available form for growth and development. Because bacteria cannot take their food in solid form, they must receive their nutrients in some kind of water solution. This solution is described as *water activity* (A$_w$) which means the amount of water available for favorable growth. Solutes (salts and sugars), as well as drying, decrease the available water and can reduce microbial growth rates.

A$_w$ is the ratio of the vapor pressure of a food to that of pure water. Water, in this case, has a value of 1.0. Typical water activity limits microbiological growth. Values for some foods are presented in the following tables.

WATER ACTIVITY LIMITS FOR GROWTH	
Group	**Minimal A$_w$ Value**
Most bacteria	0.91
Most yeast	0.88
Molds	0.82

WATER ACTIVITY OF SELECTED FOODS	
Food	**Water Activity**
Fresh fruits	.91–1.00
Pudding	.91–.99
Bread	.96–.97
Cheese	.95–1.00
Fresh meat	.95–1.00
Cakes	.90–.94
Cured meat	.87–.95
Jam	.75–.80
Honey	.54–.75
Dried fruit	.55–.80
Chocolate candy	.55–.80
Caramels	.60–.65
Dried milk	.20
Dried vegetables	.20
Crackers	.10

A knowledge of these values and their effects on microbial growth can help a foodservice operator to make judgments about the hazards of certain procedures. Information about pH and A$_w$ helps in understanding which steps are critical.

Indicator Bacteria*

By controlling the growth patterns of the indicator bacteria, the rest of the foodborne illness-causing bacterial growth is controlled. Food safety standards are established on these three processes and these indicator organisms:

1. Refrigeration (35°F to 100°F)—*Listeria*
2. Cooling (60°F to 127.5°F)—*C. perfringens*
3. Pasteurization (130°F to 165°F)—*Salmonella*

FACT SHEET

Listeria

OPTIMAL GROWTH POTENTIAL

(average generation time)

35°F	2.0 Days
40°F	1.0 Days
50°F	8.7 Hours
70°F	1.8 Hours
100°F	50 Minutes
114°F	No growth

New refrigeration code for products will be 41°F or lower.

Listeria monocytogenes is quite hardy and resists the effects of freezing, drying, and heat remarkably well for a bacterium that does not form spores. It has the ability to grow at temperatures as low as 37.4°F. The infective dose is unknown, but is believed to vary with the strain and the susceptibility of the victim. In immune-compromised individuals it is safe to assume that fewer than 1,000 total organisms may cause disease. Avoid prolonged refrigeration of products above 35°F. Follow "Use by" and "Sell by" dates.

---

*Adapted with permission from *Managing a Food Safety System* seminar. Copyright © 1992 by the Education Foundation of the National Restaurant Association.

FACT SHEET

C. perfringens

OPTIMAL GROWTH POTENTIAL

(average generation time)

50°F	No growth
60°F	6.0 Hours
70°F	2.0 Hours
90°F	23 Minutes
115°F	8 Minutes
122°F	3.0 Hours

New cooling code allows 6 hours for cooling.

140°F to 70°F	in 2 Hours
70°F to 41°F	in 4 Hours

C. perfringens forms a spore (protective shell) when growth conditions become unfavorable such as through the cooling and reheating process. In the vegetative state it produces a toxin that is not destroyed by heat and grows rapidly in dense products such as stews. The infective dose is caused by ingestion of large numbers of organisms. Rapid cooling and reheating are essential to control the growth of this organism. Meats, meat products, refried beans, sauces, soups, and gravies are most often implicated.

FACT SHEET

Salmonella

PASTEURIZATION STANDARDS

Temperature °F	Salmonella 10^7
130°F	121 Minutes
140°F	12 Minutes
150°F	1.2 Minutes
160°F	7 Seconds
165°F	3 Seconds

New code for cooking poultry and reheated foods is 165°F for 15 seconds

Salmonella is the most heat resistant of the vegetative foodborne pathogens. This chart illustrates the time and temperature required to kill 10,000,000 (10^7) organisms per gram. The pasteurization is based on standards to kill salmonella to acceptable levels. The infective dose for salmonella can be as few as 15 to 20 cells, depending on the age and health of the host and the strain of salmonella. It is estimated that 2 to 4 million cases of salmonellosis occur in the United States annually.

Disease Cycles and Infective Dose

Food contamination and foodborne illness occurs when pathogenic microorganisms multiply. An infective dose is defined as one containing the number of organisms needed to make a person sick. A large number of salmonella organisms are needed for an infective dose. Very few organisms are needed for an infective dose of shigella.

Some organisms must grow in the food after contamination in order to reach infective doses, while others simply need to be carried on the food in small numbers. Some organisms, *Staphylococcus aureus*, for example, must almost always have the conditions of time and temperature for rapid growth in a food before the dose (cells and toxin) reaches a level that produces illness in healthy adults. Other organisms, Hepatitis A virus, for example, need only be present and use food as a carrier from the ineffective host to another individual.

Knowing these differences and the sources of organisms can help in making sound decisions about food flow and risk.

Health Effects

The unpleasant symptoms of a "simple" case of foodborne illness may require absence from work, school, or leisure activities while the illness runs its course.

However, consequences to a person's health can be more severe. Diarrhea and resulting dehydration may require hospitalization, and diarrhea can lead to temporary or permanent arthritic conditions in some people. Bacteria can invade the blood (septicemia) or the membranes of the brain and spinal cord (meningitis). At worst, the human costs include death and grief.

Some people are more vulnerable than others. The very young and the very old are generally most at risk. Others at high risk include those with underlying health problems and those who are malnourished. For certain types of infections, chronic antibiotic use or pregnancy may be a risk factor.

A strong immune system plays an important role in limiting the progression of illness. Infants have incompletely developed immune systems. HIV-positive, cancer, and kidney patients are among those with suppressed immune systems. The number of Americans in these high-risk categories is increasing, and preventing "preventable" foodborne illness has become imperative.

Control of Microorganisms

By knowing the characteristics of certain microorganisms we can understand how to control them in the food flow.

Vegetative cells (V) are destroyed by cooking to the proper "kill" temperature.

Spore formers (S) are controlled with rapid cooling to 40°F and rapid reheating to 165°F, and by keeping food out of the danger zone during holding periods.

Toxins (T) Production is controlled by keeping food out of the danger zone, by storing food at 40°F or below or holding at 140°F or higher, and by rapidly cooling and reheating food.

Toxin-mediated infections are controlled by keeping food out of the danger zone and through personal hygiene and sanitation practices.

Viruses are controlled in cold and prepared foods by practicing good personal hygiene and by cooking food to its appropriate temperatures.

With all bacteria, another layer of control is to practice good personal hygiene and to avoid cross-contamination.

In summary, it is always safest to use all of the following control methods to manage for these microorganisms' characteristics:

- Cooking to proper temperatures
- Keeping food out of the danger zone
- Rapidly cooling and reheating food
- Following good sanitation and personal hygiene practices

The chart on the following page illustrates the characteristics of some common microorganisms.

BACTERIA   VIRUSES*

INTOXICATIONS

B. Cereus (V,S,T)   Hepatitis A
*C. perfringens*** (V,S)   Norwalk
Staphylococcus (V,T)
*E. coli** (V)

INFECTIONS

Salmonella (V)
Campylobacter (V)
Shigella (V)

V – Vegetative cells (cells destroyed by heating to appropriate temperature).

S – Spore former (forms protective shell when heated or cooled to protect itself).

T – Indicates heat-stable toxin.

**– Toxin-mediated infection; once ingested in large numbers, the organisms produce toxins in the intestines.

Denver Department of Public Health/Consumer Protection.

Bacterial Growth

Assume that a food contains 1,000 organisms per gram (not an uncommon finding if cooking is inadequate or if cross-contamination has occurred). Assume ideal growth conditions.

Time	Number of Organisms/gm
1 hour later	4,000
2 hours	16,000
3 hours	64,000
4 hours	256,000
5 hours	1,024,000

Temperature Effect on Bacterial Multiplication

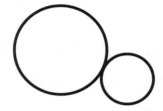

Every Half Hour   at 90°F

Every Hour   at 70°F

Every 2 Hours   at 60°F

Every 3 Hours   at 50°F

Every 6 Hours   at 40°F

Every 20 Hours   at 32°F

Every 60 Hours   at 28°F

Factors Needed for Bacterial Growth

Food (high protein)

Temperature (40°F–140°F)

Time (reproduce every 20 min.)

Moisture ($A_w > 0.85$)

Acidity (pH of 4.6 to 7.0)

Oxygen (presence or absence)

The Chain of Infection

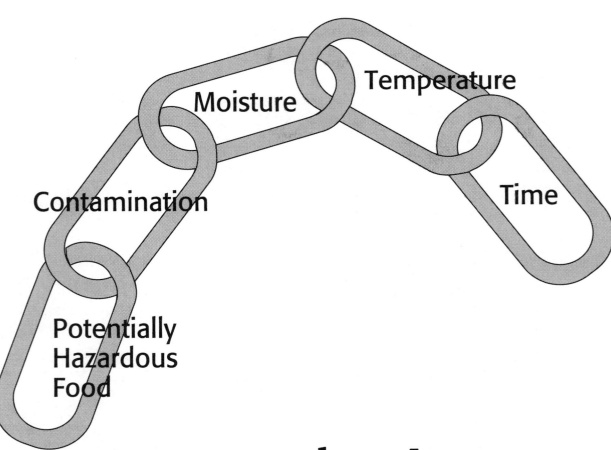

Moisture

Temperature

Contamination

Time

Potentially
Hazardous
Food

Break It!

BACTERIA THAT CAUSE FOODBORNE ILLNESS

Bacteria	Source	Disease	Control
Salmonella (infection) (13%)	Eggs, meats, chickens, bones	Salmonellosis (6–72 hrs)	Poultry 165°F, eggs 145°F, pasteurized eggs, hand washing
Staphylococcus aureus (toxins) (26%)	Skin, hair, hands, nose, open sores	Staphylococcus (1–8 hrs)	Wash hands, avoid touching face, nose, gloves w/sores, time/temp
Clostridium perfringens (produces heat-resistant spore) (16%)	Soil, dust, raw fruit/vegetables, intestines	Perfringens (8–22 hrs)	Wash hands, wash produce
Clostridium botulinum	Improperly canned foods, vacuum-packaged foods, garlic in oil, grilled onions, leftover potatoes and stews	Botulism (8–36 hrs)	Wash hands, time/temp
Shigella	Fecally (1–7 days) contaminated food or water	Shigellosis	Wash hands, proper sewage disposal, refrigeration
Hepatitis A virus	Fecally contaminated food or water	Hepatitis (15–50 days)	Wash hands, pure water, adequate cooking of foods

Emerging Pathogens

The following disease-causing agents have been increasingly identified as causing foodborne disease:

Campylobacter jejuni
E. coli 0157:57
Listeria monocytogenes
Salmonella enteriditis
Norwalk virus

These pathogens are summarized in the table on page 249 with their sources, symptoms, and control measures. Factors in the "emergence" of pathogens include:

- *Changes in eating habits*, such as eating lightly cooked foods, greater availability of foods at every corner and in the global food market.
- *Awareness of hazards and risks*, such as viruses from contaminated waters; also public awareness because of increased frequency of reported incidents.
- *Demographic changes*—population shifts from rural to higher-density urban centers, a greater number of immune-compromised individuals, aging populations, and more frequent travel between countries.
- *New food technology,* such as new-generation packaged foods.
- *Handling practices,* such as keeping eggs at room temperature, time/temperature abuse of new-generation packaged foods, neglect in properly breaking down to clean and sanitize production equipment.
- *Microorganism behavior*—evolution into new strains, as in *E. coli* 0157:H7 and *Salmonella enteriditis,* and pathogens adapting to lower temperatures.

EMERGING PATHOGENS

Bacteria	Source	Symptoms	Control
Campylobacter jejuni	Raw or undercooked meat, poultry, or shellfish, unpasteurized milk, untreated water, infected pets.	Fever, headache, abdominal cramping, bloody stools. Appears 2–5 days.	Wash hands, thoroughly cook foods; avoid unpasteurized milk, raw or undercooked meats.
Escherichia coli 0157:H7	Fecally contaminated water, raw or rare ground beef, unpasteurized milk.	Abdominal cramps, diarrhea (often bloody), nausea, vomiting — leading cause of acute failure in children. Appears 12–72 hours.	Wash hands, thorough cooking to 155°F and reheating, refrigeration below 40°F.
Listeria monocytogenes	Human and animal intestines, soil, milk, leafy vegetables, food processing. Grows slowly at refrigeration temperatures.	Disease is rare, flulike symptoms; *potentially fatal* in immune-compromised individuals; meningitis. Appears 1 day– 3 weeks.	Avoid raw milk, cheese from un-pasteurized milk. High-risk groups follow "Keep Refrigerated," "Sell By," and "Use By" labels. Thoroughly reheat frozen and refrigerated products.
Salmonella enteriditis	Raw eggs, foods containing inadequately processed eggs.	Nausea, fever, head-ache, abdominal cramps, diarrhea, sometimes vomiting. Appears 6–72 hours.	Wash hands, sanitary handling of food and equipment, adequate cooking, prompt refriger-ation.
Norwalk virus	Human carriers, fecal contamination, contaminated water, ice, shellfish, uncooked foods.	Nausea, vomiting, diarrhea, abdominal pain, headache, low-grade fever. Appears 24–48 hours.	Sanitary handling of foods, thorough cooking of foods, potable water, certified shellfish suppliers.

Who, Why, When, and Where of Foodborne Illness (and What to Do About Them)

No one would intentionally put poison in his or her food. But sometimes food becomes toxic because of lack of care in processing or because people fail to take steps to prevent foodborne illness. This happens because people often are not aware of the variety of ways foods can become contaminated.

The chart on page 251 identifies some of the more common organisms that cause food poisoning, the name of the illness each causes, conditions under which the organisms thrive in food, symptoms of illness, and the methods of preventing food contamination.

Much of the illness classified under the heading "food poisoning" involves the gastrointestinal tract (with one notable exception: botulism causes neurotoxic symptoms involving the respiratory tract). Gastrointestinal symptoms include nausea, vomiting, and diarrhea, and are sometimes known as "gastroenteritis." Although the symptoms are all somewhat similar, the organisms or their by-products that cause the illness are quite different. Those listed in the chart include bacteria, bacterial toxins, viruses, and fish toxins.

The events that result in food poisoning are also different. For instance, some bacteria, such as *Shigella,* cause illness when a small number of organisms are eaten in food such as potato salad. Because the bacteria multiply rapidly in certain prepared foods left at room temperature, prompt and proper refrigeration of the susceptible foods is the only way to prevent illness.

On the other hand, *Clostridium botulinum* bacteria are perfectly harmless to most people if eaten in foods that have not been preserved, because they are in the spore, or inactive, stage that survives when there is oxygen in the environment. Some infants, however, can become ill from eating botulinum spores. The spores remain inactive unless they are put into an anaerobic (oxygen-less), low-acid environment; for example, a can of green beans. If spores are not killed during the canning process by application of heat under high pressure, they will reactivate and resume their life cycle, or reproductive process. During the reproductive process, the botulinum toxin is produced. It is this toxin that makes the food poisonous. Contamination of preserved foods may be prevented by following the recommended established canning or other preservation methods.

Viruses have their source in the human intestinal tract and can get into food directly from handling by a person who fails to wash his or her hands after using the rest room. The more personal handling of a food, the greater the chance of contamination and resulting illness. To control viruses, it is essential to practice proper sanitizing, hand washing, and personal hygiene habits during food preparation.

Illnesses caused by foodborne poisons and poisonous organisms can vary in intensity. The malady can be so mild that it is barely noticed or is passed off as an upset stomach or flu. Again, a lengthy hospitalization might result and, in some cases, the illness can kill. But regardless of its intensity, food poisoning is unpleasant. It is also something that can be avoided.

Edited from information compiled by Carol L. Ballentine and Michael L. Herndon of FDA's public affairs staff. *FDA Consumer,* July–August 1982.

MAJOR BACTERIAL FOODBORNE ILLNESSES

Disease and Organism That Cause It	Source of Illness	Symptoms	Prevention Methods
Salmonellosis *Salmonella* (facultative bacteria) (bacteria; more than 2000 kinds) Infection	May be found in poultry, eggs, meats, fish, milk, and products made with them. Multiplies rapidly at room temperature.	Onset 6–72 hours. Lasts 2–3 days. Nausea, fever, headache, abdominal cramps, diarrhea, and sometimes vomiting. Can be fatal in infants and in elderly and immune-compromised individuals.	Avoid cross-contamination of foods, ensure thorough cooking of foods, prompt and proper refrigeration of foods, practice good personal hygiene.
Shigellosis *Shigella* (facultative bacteria) Infection	Found in mixed and moist foods, salads, lettuce, milk, and dairy products. Food becomes contaminated when a human carrier with poor personal hygiene handles liquid or moist food that is not cooked thoroughly. Organism multiplies rapidly at room temperature.	Onset 1–7 days. Lasts 1–3 days. Abdominal pain, cramps, diarrhea, fever, vomiting, and blood, pus, or mucus in stools. Can be serious in infants and in elderly and immune-compromised individuals.	Avoid cross-contamination of foods, avoid fecal contamination by food handlers, practice good personal hygiene, use sanitary food and water sources, ensure control of flies, cool foods rapidly.
Campylobacteriosis *Campylobacter jejuni* (reduced oxygen) Infection	Bacteria found on poultry, cattle, and sheep and can contaminate the meat and milk of these animals. Chief food sources: raw poultry, meat, and unpasteurized milk	Onset 2–5 days. Lasts 7–10 days; relapses are not uncommon. Diarrhea, fever, abdominal pain, nausea, headache, muscle pain, and sometimes bloody stools.	Cook foods thoroughly, avoid cross-contamination of foods, avoid unpasteurized milk.
Listeriosis *Listeria monocytogenes* (reduced oxygen) Infection	Found in animal intestines, soil, milk, leafy vegetables, poultry, meats, seafood, and prepared, chilled, ready-to-eat foods. Grows slowly at refrigeration temperatures.	Onset 1 day to 3 weeks. Duration is indefinite, depends on treatment. Has a high fatality in the immune-compromised individuals. Flu-like symptoms: nausea, vomiting, headache, fever, backache, respiratory distress, meningitis.	Avoid raw milk, cheese from unpasteurized milk, cook foods to proper temperatures, avoid cross-contamination, clean and disinfect. High-risk groups follow "Keep refrigerated," "Sell by" and "Use by" dates on sous-vide foods.
Botulism *Clostridium botulinum* (anaerobic) Intoxication Spore Former	Common in soil and water. The bacteria produce a toxin in oxygen-free and low-acid environment. Found in canned low-acid foods, garlic-in-oil, sauteed onions, leftover potatoes, stews, meat and poultry loaves.	Onset 8–36 hours. Lasts several days to a year. Neurotoxic symptoms, including double vision, inability to swallow, speech difficulty, and progressive paralysis of respiratory system. OBTAIN MEDICAL HELP IMMEDIATELY—CAN BE FATAL.	Avoid home canned products, purchase garlic-in-oil in small quantities and refrigerate, sauté onions to order, practice time and temperature controls for sous vide items and large bulky foods, cool leftovers rapidly. Avoid using bulged canned goods.

251

MAJOR BACTERIAL FOODBORNE ILLNESSES

Disease and Organism That Cause It	Source of Illness	Symptoms	Prevention Methods
Staphylococcus *Staphylococcus aureus* (facultative) Intoxication	The toxin is produced when contaminated food is left too long in the temperature danger zone. Bacteria are found in human skin, nose, throat, infected sores, and in animals. Grows well in meats and protein foods, leftovers, salads, and cream fillings.	Onset 1–8 hours. Lasts 24–48 hours. Mimics flu; diarrhea, vomiting, nausea, abdominal cramps, and physical exhaustion. Rarely fatal.	Avoid contamination from hands, exclude food handlers with skin infections from preparation, ensure proper refrigeration, rapid cooling of prepared foods.
Clostridium Perfringens Enteritis *Clostridium perfringens* (anaerobic) Toxin-mediated Intoxication Spore Former	Bacteria are widespread in the environment, generally found in meat and poultry and dishes made with them. Multiply rapidly at room temperature.	Onset 8–22 hours (usually 12). Lasts 24 hours. Abdominal pain and diarrhea, sometimes nausea and vomiting. Symptoms last a day or less and are usually mild. Can be serious in young, elderly, and immune-compromised individuals.	Follow time and temperature controls in cooling and reheating foods. Rapidly reheat foods to 165°F.
Bacillus Cereus *Bacillus cereus* (facultative) Intoxication Spore Former	Illness may be caused by the bacteria, which are widespread in the environment, or by a toxin created by the bacteria. Found in rice and pasta dishes, spices, dry food mixes, cereal products, sauces, vegetable dishes, and salads.	Onset: ½–5 hours; 8–16 hours. Lasts 6–24 hours; 12 hours. Two types of illness: (1) abdominal pain and diarrhea and (2) nausea and vomiting.	Follow time and temperature controls: rapidly cool foods, hot hold foods above 140°F, rapidly reheat foods to 165°F.
E. Coli 0157:H7 *Escherichia coli* Toxin-mediated Infection	Found in cattle. Implicated foods are raw and undercooked ground beef, unpasteurized milk and cheeses, and fecally contaminated water and foods.	Onset 12–72 hours. Lasts 1–8 days. Severe abdominal pain, diarrhea (often bloody), nausea, vomiting. Illness is usually self-limiting, can be fatal in young, elderly, and immune-compromised individuals.	Cook ground beef thoroughly, avoid cross-contamination, practice good personal hygiene.

TERMINOLOGY

Facultative—bacteria that can grow with or without free oxygen available
Anaerobic—bacteria that grow only in the absence of free oxygen
Infection—disease that results from eating living harmful microorganisms
Intoxication—disease that results from eating toxins or poisons from bacterial or mold growth
Toxin-mediated Infection—disease that results from eating microorganisms, once ingested organisms produce toxins
Immune-compromised individual—An individual who is susceptible to becoming ill because of an existing disease or weakened physical condition

MAJOR VIRAL AND FISH TOXIN FOODBORNE ILLNESSES

Disease and Cause	Source of Illness	Symptoms	Prevention Methods
Viruses			
Hepatitis A Hepatitis A virus (HAV) Viral Infection	Chief food sources: shellfish harvested from contaminated waters. Food becomes contaminated when a human carrier with poor personal hygiene handles food that will be eaten raw.	Onset 15–50 days. Lasts 1–2 weeks in mild cases. Fever, nausea, anorexia, fatigue, followed by jaundice. May cause liver damage and death. Dark-colored urine and clay-colored stools.	Avoid fecal contamination from food handlers by practicing good personal hygiene, purchase shellfish from approved, certified sources, use sanitary water sources, cook foods thoroughly.
Norwalk Virus Gastroenteritis Norwalk and Norwalk-like viral agent Viral Infection	Transmitted by the fecal-oral route via contaminated water and foods. Water is the most common source of outbreaks. Shellfish and salad ingredients are foods most often implicated.	Onset 24–48 hours. Lasts 24–60 hours. Mild brief illness: nausea, vomiting, diarrhea, abdominal pain, headache, and low grade fever. Severe illness or hospitalization is very rare.	Avoid fecal contamination from food handlers by practicing good personal hygiene, purchase shellfish from approved, certified sources, use sanitary water sources, cook foods thoroughly.
Fish Toxins			
Ciguatera Fish Poisoning Intoxication	Certain species of tropical reef fish eat smaller reef fish that have eaten algae carrying ciguatoxin. Implicated fish include groupers, barracudas, snappers, jacks, mackerel, and trigger fish.	Onset 6 hours. Last several days in mild cases. Vomiting, itching, nausea, dizziness, hot and cold flashes, temporary blindness, and sometimes hullucinations. In severe cases neurological symptoms can last weeks to months.	The toxin is NOT destroyed by cooking. Obtain fish from an approved, certified source. Carefully select the kinds of fish served.
Scombroid Fish Poisoning Intoxication	Found most often in fish that has been allowed to decompose through time and temperature abuse, histamine is produced. Implicated fish include tunas, mahi mahi, bluefish, sardines, mackerel, amberjack, and abalone.	Onset immediate to 30 minutes. Duration is usually 3 hours, but may last several days. Flushing and sweating, a burning or peppery taste, nausea, and headache. Symptoms may include facial rash, hives, edema, diarrhea, and abdominal cramps.	Histamine is odorless, tasteless, and is NOT destroyed by cooking. Use careful time and temperature control at receiving and storage. Obtain fish from an approved, certified source.

FDA Consumer, July–August 1992.
Foodborne Pathogenic Microorganisms & Natural Toxins, Food and Drug Administration State Training Branch, 1992.
HACCP Reference Manual, The Education Foundation of the National Restaurant Association, 1993.
Denver Department of Public Health/Consumer Protection, 1994.

Hepatitis A*

WHAT IS HEPATITIS A?

Hepatitis A is an illness caused by a virus that affects the liver.

WHAT ARE THE SYMPTOMS?

Adults tend to have more symptoms and feel sicker than children; children under four years of age may not look sick or have any symptoms. The symptoms for hepatitis A are:

- fever, feeling tired
- poor appetite, nausea (sometimes vomiting)
- abdominal pain or cramping
- brownish, tea-colored urine and gray or white stools
- frequent loose stools, diarrhea
- a yellowing of the skin or the whites of the eyes—a condition known as jaundice

HOW SOON DO SYMPTOMS APPEAR?

Symptoms can appear as early as 2 weeks or as late as 6 weeks (15–50 days) after a person has been infected with the virus. For most people the symptoms appear 3–4 weeks (28 days) after being infected with hepatitis A virus.

HOW IS THE VIRUS SPREAD?

- The virus in ingested (swallowed), multiplies in the body, and is passed only in the stool (bowel movements) of the infected person—it is not in the saliva.
- The virus can be passed to others when the ill person does not wash his or her hands well after going to the bathroom (or changing diapers) and then prepares foods or touches objects (such as toys) others will put in their mouths.
- Hepatitis A is not spread by sharing drinks, kissing, hugging, or casual contact.

HOW LONG CAN AN INFECTED PERSON SPREAD HEPATITIS A?

The virus is shed in the stool as early as 2 weeks before the symptoms occur and as long as 1 week after the onset of jaundice (yellowing of the skin or whites of the eyes).

HOW CAN HEPATITIS A BE PREVENTED?

- Good hand washing is very important to control the spread of this illness. Good hand washing requires soap and warm running water. Hands should be dried with a clean towel (towels should not be shared), a paper towel, or an air dryer.
- *Wash hands:*
    Before handling food
    Before touching toys children will put in their mouths
    After going to the bathroom
    After changing a diaper or helping in toilet training
- Diaper-aged children and preschoolers need adults to help them wash their hands after each diaper change or visit to the bathroom.

Denver Department of Public Health, Consumer Protection.

WHAT SHOULD YOU DO IF YOU HAVE BEEN EXPOSED TO HEPATITIS A?

Immune globulin (if given within 2 weeks of exposure) may prevent a hepatitis A infection, or make the illness milder. This means a person exposed to hepatitis A is not as likely to become ill or pass the infection on to others.

Anyone exposed to a person with hepatitis A and is in one of the following settings should call his or her doctor or health department to receive immune globulin:

1. Child care setting (child care provider or attendee)
2. Food preparer or food handler
3. Medical setting

Hepatitis A Alerta!

PARA TODOS LOS OPERADORES DE SERVICIO DE COMIDA

Hepatitis A es una enfermedad muy seria la cual es muy facil de extenderse a otras personas y puede tener un impacto mayor en su restaurante. El año pasado, el Departamento de Salud de Denver ha visto un aumento de Hepatitis A y recientement ha visto Hepatitis A en personas que manejan alimentos. Es muy importante que todos los gerentes y personas que manejan alimentos entiendan esta enfermedad y como se extiende. Personas que manejan comida y estan infectadas con Hepatitis A pueden extander este virus a un gran numero de personas.

Que es Hepatitis A? Hepatitis A es una enfermedad causada por un virus. El virus esta presente en el excremento de la persona infectada y se puede pasar a otras personas si la persona no se lava las manos muy bién despues de que use el baño. La persona infectada entonces prepara comida que otros van a comer.

Lo que Usted Puede Hacer Para Controlar Hepatitis A:

- LAVESE LAS MANOS! Un buén lavado de manos es muy importante para controlar la existencia del virus. Discipline y observe a los empleados acerca de la higiene despues del uso del baño y antes de que maneje comida:

  1. Lavandose las manos muy bién con jabon y agua caliente;
  2. Secandose las manos con toallas de papel o secadora de aire.

- Haga un plan de acción en que los alimentos se manejen solamente con utensilios—incluyendo los condimentos.
- Los empleados deben usar guantes para preparar las comidas. Las manos deben estar bién lavadas aun si se usan los quantes para preparar los comidas.
- Nunas permita que los empleados que estan enfermos manejen comida, especialmente si tienen diarrea. Empleados que no estan muy enfermos y que pueden trabajar, podrian trabajar donde no manejen comida.

Cumpliendo con estos requisitos, usted puede proteger a sus empleados y clientes previniendo la existencia de esta enfermedad. Si usted tiene alguna duda refiriendose a la Hepatitis A. Llame al Departamento de Salud de Denver al Telefono 303-436-7330.

Signs of Hepatitis A—Development and Medical Action to Prevent Further Cases

SUNDAY	MONDAY	TUESDAY	WEDNESDAY	THURSDAY	FRIDAY	SATURDAY
Incubation period is the time between initial contact with an infectious agent and appearance of the first sign or symptom of the disease. It is 15–58 days for hepatitis A, depending upon dose; average 28–30 days.		A person who contracts hepatitis A is infectious 14 days prior to becoming ill and for a week after if jaundice occurs or 2 weeks if the symptom is absent.	Immunization to prevent contracting hepatitis A or minimize the severity of the symptoms must be done within 14 days of exposure to the virus.	1	2 Date hepatitis A virus ingested by a food-service worker	3
4	5	6	7 INCUBATION PERIOD	8	9	10
11	12	13	14	15	16	17 Date customer ate salad prepared by a foodservice worker
18	19 INFECTIOUS	20 INFECTIOUS	21 INFECTIOUS	22 INFECTIOUS	23 INFECTIOUS	24 INFECTIOUS
25 INFECTIOUS	26 INFECTIOUS	27 INFECTIOUS Date foodservice worker becomes ill	28 INFECTIOUS	29 INFECTIOUS	30 INFECTIOUS	31 INFECTIOUS Last day for customer to be immunized with gamma globulin
INFECTIOUS	INFECTIOUS	INFECTIOUS	INFECTIOUS	INFECTIOUS	INFECTIOUS	INFECTIOUS

This is only ONE example of how a person who has ingested the hepatitis A virus can be affected. Remember, the incubation period is 15–58 days. In the illustration given, the foodservice worker could have gotten ill much sooner or not until the next month.

Salmonella*

CAUSE: SALMONELLA—bacteria with many subtypes

SYMPTOMS: Diarrhea, fever, cramps, sometimes nausea and vomiting. Can be fatal in infants, the elderly, and immune-compromised individuals. Lasts 2–3 days. Incubation period 6–72 hours, average ½ to 1½ days.

SOURCE: May be found in raw meats, poultry, eggs, fish, milk, and products made with them. Multiplies rapidly at room temperature.

HOW DO YOU GET IT?

The bacteria are commonly present in poultry, sometimes in beef and pork, and may be present in unpasteurized dairy products. They are in the droppings of animals (dogs, cats, cows, sheep) infected with Salmonella. They may be in untreated stream water. They are in the feces (stools) of people who are infected with Salmonella. You get it by eating food or drinking beverages that have been contaminated with feces from people or animals that are infected with Salmonella. You can also get it if your own hands are contaminated by feces or raw poultry or meat when you eat with your hands or put your hands in your mouth. Because of this, the bacteria may be spread within the family unit and between small children. Salmonella is *not* spread by coughing or sneezing.

HOW CAN YOU KEEP FROM GETTING IT?

1. Cook poultry and meats thoroughly.
2. Take care when handling uncooked poultry or meat—don't lick your fingers, smoke a cigarette, or handle a drinking cup; wash the cutting board thoroughly before cutting anything else on it.
3. Don't drink untreated water (from stream, lake, etc.)
4. Don't drink raw milk or eat other unpasteurized dairy products.
5. Wash your hands after changing diapers.
6. Keep children away from animal droppings.

HOW CAN YOU KEEP FROM SPREADING IT IF YOU HAVE IT?

Wash your hands—after going to the bathroom, before preparing food, before wiping children's faces. Ill persons should not prepare food if it can be avoided. Those who prepare food outside the home should not return to work until their doctor tells them it's OK.

Colorado Department of Health.

Campylobacter*

CAUSE: CAMPYLOBACTER JEJUNI—bacteria

SYMPTOMS: Diarrhea (sometimes bloody), fever, cramps, tiredness, and sometimes vomiting. Lasts 7–10 days. Incubation period 1–7 days, average 3–5 days.

SOURCE: Bacteria found in poultry, cattle, and sheep and can contaminate the meat and milk of these animals. Chief food sources: raw poultry and meat and unpasteurized milk.

HOW DO YOU GET IT?

These bacteria are commonly present in poultry and may be present in unpasteurized milk. They are in the droppings of animals (dogs, cats, cows, sheep) infected with Campylobacter. They may be in untreated stream water. They are in the feces (stools) of people who are infected with Campylobacter. You get it by eating food or drinking beverages that have been contaminated with feces from people or animals that are infected with Campylobacter. You can also get it if your own hands are contaminated by feces or raw poultry when you eat with your hands or put your hands in your mouth. Because of this, the bacteria may be spread within the family unit and between small children. Campylobacter is not spread by coughing or sneezing.

HOW CAN YOU KEEP FROM GETTING IT?

1. Cook chicken and turkey thoroughly.
2. Take care when handling uncooked poultry—don't lick your fingers, smoke a cigarette, or use a drinking cup; wash the cutting board thoroughly before cutting anything else on it.
3. Don't drink untreated water (from a stream, lake, etc.)
4. Don't drink raw milk or eat other unpasteurized dairy products.
5. Wash your hands after changing diapers and before eating.
6. Wash your hands after emptying a cat litter box.
7. Keep children away from animal droppings.

HOW CAN YOU KEEP FROM SPREADING IT IF YOU HAVE IT?

Wash your hands—after going to the bathroom, before preparing food, before wiping children's faces. Ill persons should not prepare food if it can be avoided. Those who prepare food outside the home should not return to work until their doctor tells them it's OK.

*Colorado Department of Health.

Shigella

CAUSE: SHIGELLA—bacteria

SYMPTOMS: Diarrhea (sometimes bloody), fever, cramps, nausea, sometimes vomiting. Lasts 1–3 days. Incubation period is 1–7 days, average 1–3 days.

SOURCE: Food becomes contaminated when a human carrier with poor sanitary habits handles liquid or moist food that is then not cooked thoroughly. Organisms multiply in food above room temperature. Found in milk, dairy products, poultry, fruits, and vegetables.

HOW DO YOU GET IT?

The bacteria are in the feces (stools) of people who are sick with Shigella infection. You get it by eating food or drinking beverages that have been contaminated with feces from a person infected with Shigella, such as when food is prepared by an ill person with contaminated hands. You can also get it if your own hands are contaminated by feces when you eat with your hands or put your hands in your mouth. Because of this, the bacteria is spread easily within the family unit and between small children. It is not spread by coughing or sneezing.

HOW CAN YOU KEEP FROM GETTING IT?

1. Wash your hands after changing diapers.
2. Insist that everyone in your household wash his or her hands after using the toilet and before preparing food or feeding children.
3. Wash your hands before eating.

HOW CAN YOU KEEP FROM SPREADING IT IF YOU HAVE IT?

Wash your hands—after going to the bathroom, before preparing food, before wiping children's faces. Ill persons should not prepare food if it can be avoided. Those who prepare food outside the home should not return to work until their doctor tells them it's OK.

E. Coli

CAUSE: ESCHERICHIA COLI 0157:H7 — bacteria

SYMPTOMS: Severe cramping (abdominal pain) and diarrhea, which is initially watery but becomes grossly bloody. Occasionally vomiting occurs. Fever is either low-grade or absent. The illness is usually self-limiting. Lasts 1–8 days. Incubation period is 12–72 hours, an average of 2 days. Can lead to hospitalization, kidney failure, and hemolytic anemia. High fatality rate with the very young and elderly.

SOURCE: Most implicated cases involve raw or undercooked hamburger. Contamination possible during slaughtering process or during grinding process. The bacteria are in the feces (stools) of people who are sick with *E. coli* infection. Cross-contamination can occur with food handlers or infected person.

                  Virulent organism: a relatively few organisms cause infection; withstands freezing. Survives temperatures as high as 140°F and survives at very low pH.

RECENT OUTBREAKS (1992–1993):

1. Washington, California, Nevada — 500 cases, 3 deaths (children) one reason: frozen contaminated hamburgers cooked to less than 155°F.
2. Oregon — 6 cases: mayonnaise (possible cross-contamination).
3. New England — 6 or more cases: refrigerated unpasteurized cider.
4. California: raw goat's milk.

HOW CAN YOU KEEP FROM GETTING IT?

1. Cook hamburgers to at least 155°F.
2. Monitor grill temperatures. Monitor hamburger temperatures with thermocouple-type thermometer.
3. Avoid cross-contamination, use safe food and water supplies.
4. Avoid fecal contamination from food handlers by practicing good personal hygiene.

HOW CAN YOU KEEP FROM SPREADING IT IF YOU HAVE IT?

Wash your hands — after going to the bathroom, before preparing food, before wiping children's faces. Ill persons should not prepare food if it can be avoided. Those who prepare food outside the home should not return to work until their doctor tells them its OK.

Listeria

CAUSE: LISTERIA MONOCYTOGENES—bacteria

SYMPTOMS: Flulike symptoms, including persistent fever. Nausea, vomiting, and diarrhea may precede more serious forms of listeriosis. When listeric meningitis occurs, the overall mortality rate may be as high as 70%; from septicemia 50%; from perinatal/neonatal infections 80%, resulting in spontaneous abortion or stillbirth. In infections during pregnancy, the mother usually survives. Onset time is unknown but is probably 1 day to 3 weeks.

SOURCES: These bacteria have been associated with foods such as raw milk, cheese (particulary soft-ripened varieties), ice cream, raw vegetables, fermented raw-meat sausages, raw and cooked poultry, raw meats (all types), and raw and smoked fish. Their ability to grow at temperatures as low as 37.4°F permits multiplication in refrigerated foods. *Listeria monocytogenes* is quite hardy and resists effects of freezing, drying, and heat remarkably well for a bacterium that does not form spores.

HIGH RISK GROUPS:

1. Pregnant woman/fetus—perinatal and neonatal infections
2. Persons immune compromised—corticosteroids, anticancer drugs, graft suppression therapy, AIDS
3. Cancer patients—leukemia patients particularly
4. Less frequently reported—diabetic, cirrhotic, asthmatic, and ulcerative colitis patients
5. Elderly patients
6. Normally healthy people—could develop the disease, particularly if the foodstuff was heavily contaminated with the organism

HOW CAN YOU KEEP FROM GETTING IT?

1. Avoid raw milk, cheese from unpasteurized milk.
2. High-risk groups follow "Keep Refrigerated," "Sell By," and "Use By" labels.
3. Thoroughly reheat frozen and refrigerated products..

Vibrio Cholera

CAUSE: VIBRIO CHOLERAE—bacteria

SYMPTOMS: Diarrhea, abdominal cramps, and fever. Sometimes vomiting and nausea, or blood or mucus in stools. Diarrhea may be quite severe, lasting 6–7 days in some cases. Diarrhea will usually occur within 48 hours. Diarrhea usually lasts 7 days and is self-limiting. Symptoms can range from mild and uncomplicated to fatal. Incubation period is 1–3 days.

SOURCE: Fish and shellfish harvested from waters contaminated by human sewage. Consumption of raw, improperly cooked or cooked, recontaminated shellfish may lead to infection. Improper refrigeration of contaminated seafood will allow proliferation, which increases the possibility of infection.

HOW CAN YOU KEEP FROM GETTING IT?

1. Sanitary handling of foods.
2. Thorough cooking of seafood.
3. High-risk groups should avoid eating raw shellfish.
4. Proper refrigeration of fish and shellfish.

HOW CAN YOU KEEP FROM SPREADING IT IF YOU HAVE IT?

Wash your hands—after going to the bathroom, before preparing food, before wiping children's faces. Ill persons should not prepare food if it can be avoided. Those who prepare food outside the home should not return to work until their doctor tells them it's OK.

Norwalk Virus

CAUSE: NORWALK VIRUS and NORWALK-LIKE VIRUSES

SYMPTOMS: Nausea, vomiting, diarrhea, abdominal pain, headache, and low-grade fever. Lasts about 24-48 hours. Incubation period is about 24–60 hours. Severe illness or hospitalization is very rare.

SOURCES: Humans are the only known reservoir.

HOW DO YOU GET IT?

Norwalk is caused by poor personal hygiene. The virus does not multiply in food, but may be carried by air, water, or uncooked food to a human host. Heat destroys the virus. Transmission of the virus is a concern in foods that receive no heat processing after contamination. The Norwalk virus can survive freezing temperatures and chlorine sanitizing solutions.

Implicated foods include: raw vegetables fertilized by manure, cole slaw and other salads, raw shellfish, eggs, icing on baked goods, manufactured ice cubes, and frozen foods.

The estimated annual incidence of Norwalk is 181,000 cases.

HOW CAN YOU KEEP FROM GETTING IT?

1. Good personal hygiene. Wash your hands before handling food and utensils.
2. Wash your hands after changing diapers and before eating.
3. Use potable water supply for drinking and ice-making.
4. Purchase from certified shellfish suppliers.
5. Cook foods thoroughly.

HOW CAN YOU KEEP FROM SPREADING IT IF YOU HAVE IT?

Wash your hands—after going to the bathroom, before preparing food, before wiping children's faces. Ill persons should not prepare food if it can be avoided. Those who prepare food outside the home should not return to work until flulike symptoms disappear.

Crisis Management: Foodborne Illness Complaints

The objective of the HACCP Plan is to respond to complaints of foodborne illness and to lessen the effects of those complaints by:

Obtaining complete, reliable information

Evaluating the complaint

Dealing positively with regulatory agencies and the media

Reapplying HACCP to initiate corrections, prevent recurrence, reduce liability

Crisis Management

Many people believe that it will never happen to them, that because they have never had a problem before, why should they worry about it? Experts believe the question is not whether, but when a crisis will occur.

By preparing for a crisis before it happens, you will be able to respond quickly and make better decisions.

There are several types of foodservice crises:

- Foodborne illness complaints
- Foodborne illness outbreak
- Security
    Robbery
    Theft
    Break-ins
    Murders
    Bomb threats
- Natural Disasters
    Floods
    Earthquakes
    Electrical storms
- Other Crises
    Fires
    Power or water failures
    Injuries
- Product Recalls

Developing a Plan for Food Liability*

It is essential that every foodservice operation develop a plan to meet potential crises. The following is an example of a plan for a foodborne illness crisis.

Step 1: Plan and develop policies and procedures.

Crisis management is an organized and systematic effort to:

- Restrict the possibility of a likely crisis
- Manage and conclude an existing crisis
- Evaluate and learn from an incident of crisis

---

*S.A.F.E., reprinted with the permission of the National Restaurant Association.

The objectives of the plan should be to preserve human life, property, and supplies and serve to correct existing problems that led to a crisis. By planning in advance of a crisis, you will be able to respond quickly and make better decisions.

An effective plan includes:

- Identification of potential areas of crisis and steps to avoid such situations
- Management and resolution of crises
- Evaluation during and after crises

Step 2: Designate specific person or a "crisis team."

Refer all food illness complaints to the team, which should include the manager or someone with similar authority. Every employee should be trained in the policies and procedures to obtain the basic information, and instructed to refer a complaint to the team for handling.

Give the designated crisis team authority to act and to direct other employees. Make sure that all statements to complainants, regulatory authorities, or the media come only from the team, so that you don't give out inconsistent information.

Step 3: Take the complaint and obtain information. Use standardized forms and procedures.

Whoever takes the complaint should follow these guidelines:

Get all pertinent information that is possible, without "pressuring" the complainant. Use standardized forms to avoid omitting any information. See the sample complaint report on page 271.

Remain polite and concerned. Use your interpersonal skills. Don't argue, but don't admit liability. For example, you might say, "I'm sorry you are not feeling well," *not* "I'm sorry our food made you sick." Also, don't offer to pay medical bills or other costs, except on the advice of your attorney or insurance agent.

Let the person tell his or her own story—don't introduce symptoms. (People tend to be very suggestible about illness, and if you suggest a symptom, they may report that symptom, because they think they should.) Just record what the person tells you.

Note the time the symptoms started. This is very helpful in identifying the disease and can work to clear your involvement.

Try to get a food history. Most people blame illness on the last meal they ate, but many diseases have longer incubation periods. Again, this could clear your involvement. (However, don't press—most of us can't remember what we ate for more than a few hours.) If applicable, try to include food eaten before and after the person was in your establishment.

Don't play doctor! Avoid the temptation to interpret symptoms or advise on treatment. Simply gather the information, remain polite and concerned, and tell the complainant you will be back in touch with him or her. Then do call back when you have something to report.

Step 4: Evaluate the complaint.

The next step will be to evaluate the complaint, so that you can handle it and respond appropriately:

Evaluate the complainant's attitude.

Resist the urge to argue or to "pay off" the complainant.

Evaluate data to determine whether the complainant is describing a legitimate illness.

You will need to use your interpersonal skills to evaluate the complaint. What is the complainant's attitude? If belligerent and demanding, it may be a bluff, or it could reflect a sincere feeling that you have caused damage. Resist the urge to argue or "pay off" immediately. Note the facts on your standardized form.

Don't diagnose; however, examine the reported information for consistency.

Did the complainant eat all of the implicated serving or just a bite? Severity and duration of illness is often dose-related.

Did anyone else in the party have the same food?

Did symptoms occur immediately? Most illnesses require several hours' incubation.

Compare the complainant's implicated food(s) with other meals served during the same time period. Was anything different about the complainant's meal? Was this the only complaint?

If this complaint is an isolated case, follow your firm's policies regarding small tokens (meal coupons, etc.) to soothe the customer and win back goodwill. Be careful to avoid admitting liability. Even in an isolated case, it would be good to review your processes and records for any possible weak spots, make needed corrections, then file the complaint form for future reference.

If your evaluation indicates that the complaint is valid, obtain outside help for the rest of your investigation. Contact your attorney, insurance agent, and the health department promptly.

Whether you decide to handle the complaint privately or get involved with regulators and outside advisors, you should investigate the complaint by reapplying the principles of HACCP. First, flow chart the implicated food. Determine whether critical control points were properly controlled. See the sample Food Product Flow Chart on page 272. Next, review the information and your records including the following:

Menus and relevant forms and logs

Numbers of the implicated meals served, and other complaints

Recent changes in suppliers, employees, process, volume

Correctly operating equipment

Recent employee illnesses (before and after the implicated meal)

Any indication that requirements for critical control points were not met.

If you still have the implicated item(s), remove it from sale and isolate it, wrapped securely and marked, "Do Not Use." If the item is frozen, it should remain frozen. Otherwise, refrigerate it pending further instructions from the laboratory. Consult with your attorney about the effects of samples on liability. Samples could clear you, but they could also confirm diagnosis.

The health department may request its own samples. If it does, you should take duplicate samples and arrange to analyze separately for comparison. For your samples, it is preferable to contract with a private laboratory. Remember, sampling and analysis are "educated guessing" and may or may not prove anything.

Use all the combined information to make any needed changes in your operation to prevent recurrence (or use the data to clear your establishment). Resolve the complaint in accordance with your policies and advice from your attorney and insurance agent about giving any coupons or payments. Keep all complaints filed and indexed for reference.

Dealing with the Health Department

Should you call the health department? This is a difficult question and, like other management decisions, demands that you apply good judgment. It might seem ridiculous to invite regulatory criticism. On the other hand, some local ordinances may require you to report suspected outbreaks of illness.

The safest course is generally to call the health department if you believe the complaint is valid, and certainly if there are multiple complaints or hospitalizations involved. The health department will eventually be notified, and it is to your benefit to establish a cooperative stance (even better to establish cooperation before the problem appears).

Moreover, a health department inspection of your operation can clear you and protect against frivolous or fraudulent claims. You should consult with your attorney about the specific laws in your jurisdiction concerning your rights and responsibilities. Generally, however, a health department is authorized to:

- Take reasonable samples of suspect foods
- Prevent the sale of suspect foods
- Require medical and laboratory examinations of employees
- Exclude suspect employees from food handling duties
- In extreme cases, order the facility to be closed

Develop a plan to deal effectively with the health department or other regulatory agencies. Depending on the situation and your interaction, these agencies have the potential to help you and to minimize negative effects of an incident (including publicity), or to harm your operation by overreacting or making strong statements to the media.

An inspection by health department investigators may help to clear you in a dispute. Be frank and candid; don't get caught in a "coverup."

- Be cooperative.
- Make available appropriate records for review (customer charge slips and dealer invoices).
- Allow investigators reasonable access to observe whatever they request.

Dealing with the Media

Like it or not, a report of foodborne illness can be newsworthy. You should be ready to deal with the media.

Be cautiously cooperative. Have your facts straight before you speak; answer simply, without jargon, and avoid creating a controversy that could make the story more interesting. Remain as professional as possible, stay calm, don't allow yourself to be provoked. Keep your answers positive, not defensive. It is possible to use the media to your advantage, to tell your side in a favorable light.

One planning technique suggests that you imagine your worst nightmares being described in the newspapers or on television, then practice answering those potentially embarrassing questions until you feel and act comfortable.

Stick to the truth and don't try to bluff; if you don't know the answer, say so and arrange to get back with the information later. Also, remember that "no comment" tends to sound like an admission of guilt.

> *I*f a little knowledge is dangerous, where is the man who has so much as to be out of danger?
>
> *–Author Unknown*

Foodborne Illness Complaints
Crisis Management

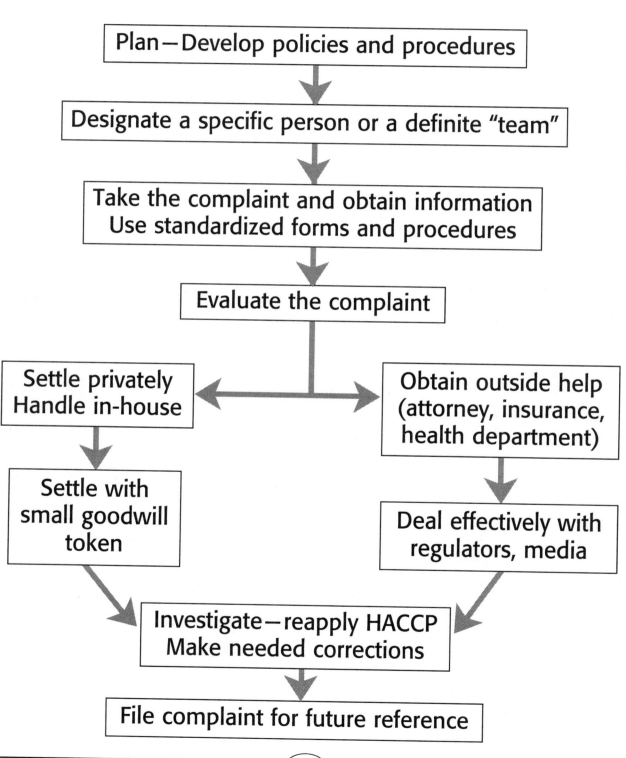

Plan—Develop policies and procedures

Designate a specific person or a definite "team"

Take the complaint and obtain information
Use standardized forms and procedures

Evaluate the complaint

Settle privately
Handle in-house

Obtain outside help
(attorney, insurance,
health department)

Settle with
small goodwill
token

Deal effectively with
regulators, media

Investigate—reapply HACCP
Make needed corrections

File complaint for future reference

Sample Foodborne Illness/Complaint Report*

Illness or food complaint report procedures for restaurant employees

1. If the manager and/or supervisor is present when a complainant calls, refer the complainant immediately. If not present, get information on form, refer to manager, assure caller manager or supervisor will call back.

2. Be polite and concerned—*Do not argue.*
3. Obtain all applicable information on form.
4. Do not admit responsibility.
5. Do not diagnose or advise on treatment. Do not suggest symptoms.

Complainant name: _____ Phone (work) _____

Address: _____ Phone (home) _____

_____

Others in party? _____ Get names and addresses

_____ (use back of form if

_____ additional space is

_____ needed)

Onset of symptoms: Date: _____ Time: _____

Symptoms: ☐ Nausea ☐ Diarrhea ☐ Fever ☐ Blurred vision

☐ Vomiting ☐ Dizziness ☐ Headache ☐ Abdominal cramps

Other: _____

Medical Treatment: Doctor _____

(Hospital) Name _____ Address _____ Phone _____

Suspect meal: _____

Location: _____

Time and Date: _____

Identification (brand name, lot number) _____

Description of meal: _____

Leftovers: _____ (Refrigerate, do not freeze)

	Date:	Time:	Location:	Description:
Other foods or beverages consumed before or after suspect meal:				

Other agencies notified?	Agency	Person to contact	Phone

Remarks: _____

Report received by: _____ Date: _____ Time: _____

Referred to: _____

*S.A.F.E., reprinted with the permission of the National Restaurant Association.

271

Food Product Flow Chart*

Establishment_____ Date_____

Address_____

Food Item_____

*Mark the critical control points with CCP.

Inspected by_____ Received by_____

*Denver Public Health Department, Consumer Protection.

272

Preliminary Foodborne Illness Investigation*

RESTAURANT NAME: _____

ADDRESS: _____

SUSPECTED FOOD/BEVERAGES (be specific) _____

DATE MEAL EATEN: _____ TIME: _____ ONSET DATE: _____ ONSET TIME: _____

CALLERS NAME: _____ PHONE: _____ INCUB TIME: _____

ADDRESS: _____

PERSON'S NAME WHO BECAME ILL: NAME / AGE / SEX / OCCUPATION

_____

DOCTOR SEEN: ☐ YES   ☐ NO   DIAGNOSIS/ LAB RESULTS: _____

CLINIC NAME / DOCTOR'S NAME: _____

ADDRESS: _____ PHONE: _____

        ☐ STOOL   ☐ BLOOD   OTHER _____ RESULTS: _____

DATE RECEIVED: _____ CALL RECEIVED BY: _____

SYMPTOMS

☐ VOMITING	☐ FEVER	☐ NAUSEA	☐ BURNING MOUTH
____ NO. OF DAYS	☐ CHILLS	☐ MUSCLE ACHE	☐ ITCHING
____ NO. OF TIMES	☐ CRAMPS	☐ EXCESS SALIVATION	☐ RASH
☐ DIARRHEA	☐ HEADACHE	☐ COUGH	☐ DIZZINESS
____ NO. DAYS	☐ PERSPIRATION	☐ METALLIC TASTE	☐ NUMBNESS
____ NO. OF TIMES			☐ DOUBLE VISION
☐ BLOODY	OTHER _____		
☐ EXPLOSIVE	_____		
☐ WATERY			

FOOD HISTORY

1ST 24 HOURS/DATE MEAL CONSUMED	ALL FOODS CONSUMED	WHERE
DINNER		
LUNCH		
BREAKFAST		

2nd 24 HOURS (PREVIOUS DAY)

DINNER

LUNCH

BREAKFAST

3rd 24 HOURS (2 DAYS PRIOR)   ALL FOODS CONSUMED AT RESTAURANTS OR FROM CATERERS

DINNER

LUNCH

BREAKFAST

1. Was this a take-out order?
2. (If yes) Elasped time between pickup to consumption _____ HRS
3. Are there any other ill contacts (including pets)?_____
4. If yes to #4, please list symptoms _____
5. Please note anything unusual noticed about meal? (temperature, taste, etc.)_____

(PLEASE USE REVERSE SIDE IF MORE ROOM IS NEEDED)

*Denver Public Health Department, Consumer Protection.

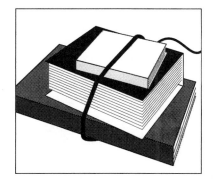

Employee Training Materials

C herish your
vision and
your dream
as they are the
children of
your soul;
the blueprints of
your ultimate
achievements.

–Napoleon Hill

Employee Hiring and Training

Effective protection and safety for customers begins with an organized approach to the hiring, training, and supervising of employees. When properly trained, employees recognize food safety hazards and apply management's operating policies and procedures.

- Write food safety requirements into job descriptions.
- Ask relevant food safety questions in interviews.
- Employ applicants who meet the job qualifications.
- Give employees a written copy of personal hygiene and safe food-handling rules and enforce them consistently. See the sample food safety procedure sheets on pages 277–278.
- Present management's dedication to food safety, and support the employees' commitment to food safety.
- Incorporate food safety and sanitation training into task learning.
- Monitor supervision of food safety and hand washing practices.
- Evaluate training needs for food safety and sanitation.
- Revise practices and procedures as job requirements change and the science of food safety evolves.
- Assess the proficiency of employee training.
- Evaluate employee performance on meeting food safety standards.

Food Safety Procedures*

PERSONAL HYGIENE

Personal hygiene is important because poor practices, such as not washing hands properly or coughing over food, may result in customers' becoming ill.

GROOMING

Bathe daily, use deodorants, change into clean clothes daily. Wear proper work shoes (no opened-toed shoes) and keep them clean.

Keep fingernails clean and trimmed. False fingernails/nail polish should not be worn to work.

Jewelry is not to be worn, as it gets dirty, can get lost in food, or can even cause injury when caught by hot or sharp objects or equipment.

CLOTHES

Wear clean clothes in good repair. If no specific uniform is required, wear plain clothes (with no writing or designs). The minimum dress requirement for each position is described below:

Wait Staff — White wing collar tux shirt, black banded bow tie, black pants or skirt, black shoes (men — black socks, women — hose).

Dish Washers — White shirt, dark pants. Hat style per management approval.

Busser/Line Server — White shirt, dark pants; clean, neat and pressed appearance. Hat style per management approval.

Cook — White oxford-style shirt or chef coat. White or checked pants (dark pants per management approval). Hat style per management approval.

Chef or Carver — Chef coat, white or checked pants, and chef hat.

HAIR RESTRAINTS

Hats and hairnets are considered proper hair restraints. Hair restraints are required to keep hair and its contaminants out of food. After touching hair or face, follow hand-washing procedures.

APRONS

Wear a clean apron. An apron should not be used as a hand towel. Follow hand-washing procedures after touching or wiping hands on an apron. Aprons should be removed when leaving the food preparation area.

ILLNESS

Anyone who is sick with flu-like symptoms should not work with food. Inform your supervisor if you have a severe cold or diarrhea.

CUTS, ABRASIONS, AND BURNS

Wounds should be bandaged. Cover bandages with waterproof protector such as rubber or plastic gloves. Inform supervisor of all wounds.

*Hospitality Personnel Services, *Temp Staff Handbook.*

HAND WASHING

Thoroughly wash hands and exposed portions of your arms with soap and warm water after touching anything that can be a source of contamination:

 Before starting work and during work as often as needed to keep clean.

 After break times, eating, or drinking.

 After personal activities, such as smoking.

 After sneezing, coughing, or using the toilet.

 After touching raw foods, meats, shell eggs, or fresh produce.

 After handling dirty dishes, utensils, or equipment.

 After handling trash, sweeping, or picking up items from the floor.

 After using cleaners or chemicals.

 After touching any other sources of contamination, such as a phone, money, door handles, or soiled linens.

FOLLOW PROPER HAND-WASHING STEPS

- Use soap and running water.
- Rub your hands vigorously with soap lather for 20 seconds.
- Wash all surfaces, including:
  backs of hands
  wrists
  between fingers
  under fingernails
- Use nail brush around and under fingernails.
- Rinse well under running water.
- Dry hands with paper towel.
- Turn off the water and turn door knob using a paper towel instead of bare hands.

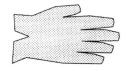

PLASTIC GLOVES

Wear plastic gloves after thoroughly washing your hands; this prevents contamination of the gloves. Change gloves often and under the same circumstances as you would wash your hands. Wash your hands after removing the gloves; this prevents contamination from the hands.

278

Safe Food Handlers*

In order to assure food safety, management also has the following responsibilities to food handlers:

- Develop job assignments.
- Provide adequate facilities.
- Lead by example.
- Provide ongoing supervision and training.

To evaluate your employee training program, ask these questions:

Did the training produce results on the job?
If intended results were not produced, why not?

Management's personnel procedures are also essential in protecting customers from food-borne illness:

- Hiring the right employees.
- Using task analysis to determine the food safety and sanitation requirements for each job.
- Establishing employee rules for personal hygiene.
- Supervising and training employees in good personal hygiene and food safety practices.

For example, see the sample task analysis and food safety objectives for one personnel category on pages 280 and 281.

*Adapted with permission from *Managing a Food Safety System* seminar. Copyright © 1992 by the Education Foundation of the National Restaurant Association.

Sample Task Analysis for a Line Cook*
(Teach by demonstration)

Personal Hygiene
- Remove all jewelry and change into a clean uniform in the employee locker room.
- Put on a hair restraint.
- Wash hands following the steps posted above the hand washing sink.

Preparation Tasks
- Thaw frozen meat products in the refrigerator below cooked or ready-to-eat foods on the lowest shelf.
- Wash hands before preparing meats and main dishes, and wash hands after handling raw product.
- Use a newly cleaned and sanitized cutting board and knife for each raw product.

Cooking Tasks
- Cook all meats and other ingredients to their recommended minimum temperatures.
- Measure temperatures with a clean and sanitized thermometer.
- Follow corrective action if food does not reach the recommended minimum temperature.
- Minimize hand contact with the cooked product, and make sure hands have been washed. Use clean and sanitized utensils.

Cleaning and Sanitizing Tasks
- Clean and sanitize small equipment.
- Keep own working area clean.

*Reprinted with permission from *HACCP Reference Book*. Copyright © 1993 by the Education Foundation of the National Restaurant Association.

Sample Food Safety Objectives of a Cook*

After food safety training, the cook will be able to:

1. Explain the fundamentals of sanitation and understand why it is necessary to learn food safety and sanitation principles and procedures.

2. Know the hazards to safe food and how they can lead to food-borne illness.

3. Practice good personal hygiene to avoid contamination of food and food-contact surfaces.

4. Follow procedures to prevent foodborne illness during food preparation, storage, and serving, using critical control points in an HACCP-based system of food safety.

5. Use steps to avoid cross-contamination during food handling duties.

6. Use a bimetallic stemmed thermometer to measure internal temperatures of food during preparation.

7. Follow all necessary steps to clean and sanitize equipment and work area.

*Adapted with permission from *Managing a Food Safety System*. Copyright © 1992 by the Education Foundation of the National Restaurant Association.

How to Get Employees Involved

As part of an organized approach to food safety, it is important to ensure the involvement of all foodservice employees. The following guidelines can be helpful:

- Develop a food safety committee.
- Establish accountability for meeting food safety responsibilities.
- Allow for employee input to identify conditions hazardous to food safety and to bring such conditions to management's attention.
- Provide training prior to all new job assignments.
- Update training at least annually as work processes and ingredients change.
- Maintain records of training.
- Train all supervisors in pertinent food safety matters.
- Schedule an available HACCP certified manager on all shifts.
- Develop procedures for problem reporting, problem investigation, corrective action, and follow-up.

Make sure all staff members realize that achieving food safety requires a team effort. The following outline lists specific measures to effect this goal:

1. Share information in staff meetings. Send members of the food safety committee to food safety seminars. Implementation works best when at least two staff members have attended HACCP seminars.

2. Set a good example—this is the best way to start remodeling behavior.

3. Make handouts from your manual for staff meetings:
   Copy of overheads
   Copy of employee handouts
   Copy on colored paper for posters

4. • Equip every work station with a thermometer. Let staff record temperatures over a period of time (cool down, reheating, and holding). Let employees see the patterns they need to change. Use copies of temperature charts included in this manual (pages 198, 200, 220).

   • Conduct a thermometer calibration demonstration (see page 86).

5. Let staff set up a cool-down experiment: Use the same quantities of product (or water) at the same temperature in three or four containers of the same size. Record the temperature every 15 minutes to 1 hour (depending on product and size of container). Test and record the cool-down temperature from these different cooling methods (see page 113).

   • Room temperature
   • Product in refrigerator
   • Ice bath, no stirring
   • Ice bath, stirring every 15 minutes
   • Plastic container covered with plastic wrap

6. Implement the use of temperature log charts for cool down, reheating, and holding. Have employees record temperatures.

7. Implement the use of temperature logs of refrigeration and hot-holding equipment. Have employees record temperatures.

8. Conduct a chemical test strip demonstration of sanitizer products.

   - Give each work area its own test strips.
   - Use the charts on pages 79, 80 in this manual for proper dilution.

9. Standardize sanitizing solutions, by using sanitizer tablets or by assigning one person to mix solutions for the whole operation each day.

10. Buy red and blue buckets for soapy water and sanitizer solution.

11. In a staff meeting:

   - Identify the potentially hazardous foods on your menu. Discuss which menu items have the most potential for risk.

   - Review critical control points. Observe those CCPs throughout the storage, preparation, holding, and serving processes.

   - Establish control procedures for the CCPs to guarantee safe food. Rewrite your recipes incorporating the controls.

12. After training, assign a team of two or three people to review recipes, noting potential problem areas. Revise these recipes according to current guidelines.

13. Appoint employees to take turns being an observer, for 10 to 15 minutes each, to monitor employee handling practices and procedures. Have employees report findings in a staff meeting. This must be done without pointing to individuals or naming names. At the conclusion of the staff meeting, ask all employees to state, individually, what they have learned and to identify one habit of their own they will work on to change.

14. Implement and post an Employee Food Safety Progress Chart. Acknowledge attendance in food safety training meetings. Recognize employees who are observed following established safety procedures.

15. Appoint a team to take slide photos of various tasks as they are performed, showing progress in achieving food safety. When the slides are shown to staff, the team can mention items needing continuing effort.

16. Set up incentive programs for food safety.

In addition, you might consider these training aids:

- Borrow videos from your local health department and/or restaurant association for your employee training meetings.
- Set up a food safety ring binder with dividers. Encourage employees to use, review, and add new information.

- Set up a food safety bulletin board. Allow employees to post cartoons, food safety tips, temperature and flow charts they have made. (Perhaps showing some items made before, after, and during training).
- Use a variety of resources to promote continued staff involvement: FDA pamphlets, equipment manufacturers, purveyors, equipment specialists, chemical purveyors, colleges, county extension agents, meat and dairy marketing associations, and the American Red Cross.

Employee Orientation

Employee orientation should include food safety standards as well as other company policies. The following checklist can be adapted to document discussions on the company's food safety standards. It may also be used to document food safety training as employees move into different positions, such as from dish washer to food preparation or grill cook.

Performance Standards

When managers incorporate food safety standards into performance evaluations, it lets employees know that food safety is their responsibility and that they will be held accountable for it. Employees must have clean personal habits, handle food safely, and practice proper sanitary procedures. The following list is an example of how food safety responsibilities may be written into performance standards.

Food Safety Checklist

Name: _____ **Date/Comments**

Has completed introduction to food safety—employee health and hygiene

_____

Demonstrates proper hand washing procedures

_____

Washes hands frequently as required using hand sink

_____

Wears clean uniform; wears hair restraint; maintains grooming standards

_____

Understands that personnel with infections will be restrained

_____

Eats and smokes in designated areas

_____

Follows safe food preparation and serving procedures

_____

Follows safe thawing procedures

_____

Stores and handles raw products properly

_____

Handles food with minimal time in danger zone

_____

Checks temps frequently; uses sanitized thermometer

_____

Follows rapid cooling procedures to 40°F

_____

Reheats foods quickly to 165°F

_____

Holds foods above 140°F or below 40°F

_____

Uses sanitizer at proper concentration

_____

Cleans and sanitizes work area and utensils

_____

Follows proper handling of utensils, dishes

_____

Sanitizes equipment/washes hands before handling and after handling raw food.

_____

Incorporating Food Safety Accountability

Food Production

Meets service and production deadlines. Meets quality standards of food and service. Follows recipes and portions. Follows *safe* food-handling practices: Sanitizes equipment between handling potentially hazardous foods. Works with food minimal time in danger zone (total time not to exceed 4 hours). Plans ahead and thaws food in refrigerator. Checks temperature with *sanitized* and calibrated thermometer.

Storage and Holding

Follows *safe* food storage, thawing, holding, cooling, and reheating standards. Rapidly cools foods to 40°F within 4–6 hours. Maintains temperature logs of cool down, reheat, and assigned equipment. Checks temperature with *sanitized* and calibrated thermometer each time.

Service Standards

Meets quality standards of foods and services. Follows proper service procedures. Maintains proper hot and cold food temperatures. Maintains temperature logs of served food. Checks temperature with *sanitized* and calibrated thermometer each time.

Receiving and Inventory

Checks in deliveries, checks for product specification, quality, condition, temperature, and invoice accuracy. Stores and handles raw products properly. Follows proper product rotation.

Training

Has attended or received updated (at least once a year) training in food safety procedures, new methods, and new ingredients to reduce risk.

Safety and Sanitation

Follows safe hand-washing and food-handling practices. Washes hands frequently. Uses sanitizer solution at proper concentration and with correct procedures. Cleans equipment per schedule. Has no critical deficiencies in his or her area of responsibility. Meets employee health, hygiene, and grooming standards. Eats and smokes in designated areas.

Training Aids

The following employee training aids are available for foodservice operators.

Slides and Transparencies

Food Safety Quality Assurance for Food Service Employees. (1992) Topics include "The Illness Hazards," "Microorganisms That Cause Illness," "Personal Hygiene," "Cleaning and Sanitizing," "Safe Food Preparation," "Proper Thermometer Care and Use," and "Correct Storage Techniques." Hospitality Institute of Technology & Management, 830 Transfer Road, St. Paul, MN 55114. Tel: 612/646-7077. Slides $1; text $15; test pack $5.

The Necessary Step—A Sanitation Package. (1982) A guide to compliance with FDA's model ordinance for sanitation in retail stores. Publications Sales, Food Marketing Institute, 800 Connecticut Avenue, NW, Washington, D.C. 20006. Tel: 202/452-8444. $75 (member), $150 (nonmember). Order no. 2-53.

Safe Food Handling: Health, an Ounce of Prevention: Serve Food, Not Illness. (1989) These 11 transparencies with instructional materials from the U.S. Department of Agriculture, Food and Nutrition Service may be obtained through interlibrary loan from the Food and Nutrition Information Center at the National Agricultural Library, Beltsville, MD 20705. Tel: 301/344-3755. F&N order no. F-322.

Your Sanitation Responsibilities. (1989) A basic 39-slide, sound cassette program for foodservice workers. Available through interlibrary loan from the Food and Nutrition Information Center at the National Agricultural Library, Beltsville, MD 20705. Tel: 301/344-3755. F&N order no. 138.

Videotapes

Back of the House II. (1987) Receiving: inspecting delivered items and monitoring food and equipment temperatures. Storage: storeroom maintenance and product inspection and rotation. (Also available in Spanish.) National Restaurant Association, The Educational Foundation, 250 S. Wacker Drive, Suite 1400, Chicago, IL 60606-5834. Tel: 800/765-2122. $99 (member), $129 (nonmember).

Back of the House III. (1987) Preparation and handling: maintaining food quality and forecasting quantities. Holding and serving: avoiding bacteria and contaminants, and labeling and dating foods. Cleaning and sanitizing: cleaning procedures and sanitizing dishes, utensils, equipment, and work surfaces. (Also available in Spanish.) National Restaurant Association, The Educational Foundation, 250 S. Wacker Drive, Suite 1400, Chicago, IL 60606-5834. Tel: 800/765-2122. $99 (member), $129 (nonmember).

Basic Facts About AIDS for Food Service Employers (tape 1). **AIDS: What You** [employees] **Need to Know** (tape 2). (1988) These two videos cover facts about the disease, laws on employment of persons with AIDS, and dealing with patron and employee fears. National Restaurant Association, 1200 17th Street, NW, Washington, D.C. 20039. Tel: 800-424-5156. $33.95.

The Danger Zone (a deli food safety and sanitation program). (1989) International Dairy-Dell Association, P.O. Box 5528, Madison, WI 53705. Tel: 608/238-7908. $65 (member), $105 (nonmember).

Food Safety Is No Mystery. (1989) A foodservice sanitation video training program produced by the U.S. Department of Agriculture, Food Safety and Inspection Service. Produced by Modern Talking Picture Service, 5000 Park Street North, St. Petersburg, FL 33709. Tel: 800/237-4599. $20.50 (English), $36 (Spanish/English—includes four posters in Spanish and English).

Foodborne Disease: It's Your Business. (1992) Introduces HACCP to foodservice owners and managers. Contact: Duain Shaw, Chief, Food Service Facilities Section, Pennsylvania Department of Environmental Resources, P.O. Box 2357, Harrisburg, PA 17120.

Grime Fighters. (1992) Covers store-level employee food handling and safety practices. Publications Sales, Food Marketing Institute, 800 Connecticut Avenue NW, Washington, DC 20006. Tel: 202/452-8444. $200 (member), $400 (nonmember).

HACCP: Safe Food Handling Techniques. (1990) Discusses how to implement an HACCP program in a foodservice operation. The only drawback for U.S. trainers is that the graphics present temperatures in degrees Celsius rather than Fahrenheit, although the narration includes both. Comes with a 20-page *Leader's Guide*. 22 minutes. Canadian Restaurant and Foodservices Association, 80 Floor Street West, Suite 1201, Toronto, Ontario, Canada M5S 2V I. Tel: 416/923-8416. $90 (members, academics, and health departments).

The Invisible Challenge: Food Safety for Food Handlers. (1989) Two-part video with four food safety posters, a probe thermometer, a copy of *Food Handler's Pocket Guide for Food Safety* and a *Consumer Guide to Food Quality & Safe Handling*. Publications Sales, Food Marketing Institute, 800 Connecticut Avenue NW, Washington, DC 20006. Tel: 202/452-8444. $95 (member), $225 (nonmember).

100 Degrees of Doom! The Time and Temperature Caper. (1988) A private-eye approach investigates the causes of a salmonella food poisoning outbreak. Includes: videocassette, instructor's guide, two posters, and a metal-stem thermometer. 14 minutes. Educational Communications Inc., 761 Fifth Avenue, King of Prussia, PA 19406. Tel: 215/337-1011. $95.

Safe Hand Washing. (1988) Covers the procedure and microbiological reasons for keeping hands clean. Comes with *Instructor's Technical Background* booklet, student lesson sheet with quiz, laminated hand-washing poster, and fingernail brush. Hospitality Institute of Technology & Management, 830 Transfer Road, Suite 35, St. Paul, MN 55114. Tel: 612/646-7077. $65 (English), $95 (Spanish/English).

Sanitation: It's Your Responsibility. Three videos on: "Preventing foodborne illness," "Keeping microbes in check," and "Personal hygiene in food service," (1989) Also available in Spanish. Advantage Media, Inc., 21356 Nordhoff Street, Suite 102, Chatsworth, CA

91311. Tel: 800/545-0166; 818/700-0504. $850/set; $395 each. These videos may also be borrowed through interlibrary loan from the Food and Nutrition Information Center at the National Agricultural Library, Beltsville, MD 20705. Tel: 301/344-3755. F&N order no. F-1787.

Sanitizing for Safety: Foodborne Illness—How You Can Prevent It. (1990) Discusses basics of correct foodservice sanitation and sanitizing with bleach. Clorox Company, Inquiry Handling Services, Receiving Department, 200 Parkside Drive, San Fernando, CA 91340.

SERVSAFE Serving Safe Food Program. Four videos: *Introduction to Food Safety: Employee Health and Hygiene, Safe Food Handling: Receiving and Store, Safe Food Handling: Preparation and Service and Cleaning,* and *Sanitizing.* (1991) Also available in Spanish. Every video includes a leader's guide. National Restaurant Association, The Education Foundation, 250 South Wacker Drive, Suite 1400, Chicago, IL 60606. Tel: 800/765-2122. $329/set or $115 each (member), $399/set or $135 each (nonmember).

The Spoilers I. (1969) Stresses using time and temperature to thwart bacterial growth, plus the importance of constantly checking for the right temperatures everywhere food is handled or stored. This tape is a basic course in food safety. Publications Sales, Food Marketing Institute, 800 Connecticut Avenue, NW, Washington, DC 20006. Tel: 202/452-8444. $50 (member), $100 (nonmember). Order no. 2-53.

Spoilers II. (1987) Covers the dangers of bacteria, methods of preventing cross-contamination, and basic steps to keep departments with perishables clean and safe. Also presents information on hot delicatessens, bakeries, and fish departments. Includes instructor's guide. Publications Sales, Food Marketing Institute, 800 Connecticut Avenue, NW, Washington, DC 20006. Tel: 202/452-8444. $50 (member), $100 (nonmember). Order no. 2-53.

The purpose of this listing is to provide a source of training aids and background information. No endorsement of named products or services is intended, nor is criticism implied of similar products or services that are not mentioned. Some of this material has not been reviewed by the authors; no statement regarding the quality or usefulness of the material is intended.

*Reprinted from *Ensuring Food Safety . . . The HACCP Way*, by Robert J. Price, Pamela D. Tom, and Kenneth E. Stevenson, Extension Service, U.S. Department of Agriculture; National Sea Grant College Program, National Oceanic and Atmospheric Administration, U. S. Department of Commerce, 1993.

Bibliography

Albertson's Service Deli #863. 1993. *Albertson's Service Deli Training Manual.* Boise, Idaho.

Boulder County Health Department, 1992. *STAR* Program. Boulder, Colo.: Boulder County Health Department.

City Market. 1994. *Food Safety/Sanitation.* Grand Junction, Colo.: City Market.

The Educational Testing Service. Center for Occupational and Professional Assessment. 1985. *Preparing to Take the Food Protection Certification Test.* Princeton, N.J.: Educational Testing Service.

Jefferson County Health Department. 1993. *Self-Test for Food Service Employees.* Jefferson County Health Department.

The National Academy of Sciences. 1991. *Seafood Training Program.* Report on Seafood Safety. Washington, D.C.: National Academy Press.

National Restaurant Association. 1991. *Make a S.A.F.E. Choice—Sanitary Assessment of Food Environment: A New Approach to Restaurant Self-Inspection.* Chicago, Ill.: National Restaurant Association.

National Restaurant Association. The Education Foundation. 1993. *The HACCP Reference Manual.* Chicago, Ill.: National Restaurant Association.

—. The Education Foundation. 1992. *Managing a Food Safety System.* Chicago. Ill.: National Restaurant Association.

—. The Education Foundation. 1993. *Serving Safe Food, Employee Guide.* Chicago, Ill.: National Restaurant Association.

—. The Education Foundation. 1991. *ServSafe.* Chicago, Ill.: National Restaurant Association.

—. The Education Foundation. 1992. *Applied Foodservice Sanitation.* Chicago, Ill.: National Restaurant Association.

Price, Robert J. August 1990. *Retail Seafood Temperature Control.* Department of Food Science Technology, Sea Grant Extension Program Publication. Davis, Calif.: University of California Press.

Pueblo County Health Department. 1992. *To Cool Potentially Hazardous Foods.*: Pueblo County Health Department.

Red Lobster. October 1992. *Quality Assurance: Solutions to Control "Critical" Food Safety and Sanitation Hazards.* Orlando, Fla.: General Mills Restaurants, Inc.

Sherer, Michael. 1993. "Anticipating Paradigm Shifts." *Food Management*, pp 93–108.

Snyder, Peter, Jr. 1992. *HACCP-Based Total Quality Management Hospital Foodservice.* St. Paul, Minn.: Hospital Institute of Technology and Management.

—. 1992. *Food Safety Through Quality Assurance Management.* St. Paul, Minn.: Hospital Institute of Technology and Management.

Taylor, Michael R. *FDA's Plans for Food Safety and HACCP: Institutionalizing a Philosophy of Prevention.* Symposium on Foodborne Microbial Pathogens, International Life Sciences Institute, Atlanta, GA. August 3, 1993.

United States Department of Agriculture. September 1990. *Preventing Foodborne Illness.* Home and Garden Bulletin No. 247. Washington, D.C.: U.S. Government Printing Office.

—. *Bacteria That Cause Foodborne Illness.* FSIS Facts. December 1990. Washington, D.C.: U.S. Government Printing Office.

United States Food and Drug Administration. Center for Food Safety and Applied Nutrition. 1992. *Foodborne Pathogenic Microorganisms and Natural Toxins.* Rockville, Md.: Food and Drug Administration, State Training Branch.

—. 1993. *HACCP Regulatory Application in Retail Food Establishments.* Rockville, Md.: Food and Drug Administration State Training Branch.

—. National Advisory Committee on Microbial Criteria for Foods. March 1992. *Hazard Analysis and Critical Control Point System.* Rockville, Md.: Food and Drug Administration State Training Branch.

United States Public Health Service. *Food Code 1993.* Washington, D.C.: Government Printing Office.

University of California Cooperative Extension. 1993. *Ensuring Food Safety—The HACCP Way: An Introduction and Resource Guide for Retail Deli Managers.* Davis, Cal.: University of California.

University of Massachusetts. Environmental Health and Safety. May 1989. *Handwashing by Food Service Personnel.* Amherst, Mass.: University of Massachusetts Press.

Glossary*

Acid   A substance with a pH of less than 7.0.

Aerobic   Able to live and reproduce only in the presence of free oxygen.

Alkali   A substance with a pH of more than 7.0.

Anaerobic   Able to live and reproduce in the absence of free oxygen.

Bacteria   Single-celled organisms, usually classified as the simplest of plants.

Biological hazard   Danger to food from disease-causing microorganisms and poisonous plants and fish.

Chemical hazard   Danger to food posed by chemical substances, especially pesticides, food additives, and toxic metals.

Clean   Free of visible soil.

Contamination   The unintended presence of harmful substances or organisms, especially in food.

Control   (a) To manage the conditions of an operation to maintain compliance with established criteria. (b) The state wherein correct procedures are being followed and criteria are being met.

Control point   Any point, step, or procedure at which biological, physical, or chemical factors can be controlled.

Corrective action   Procedures to be followed when a deviation occurs.

Criterion   A requirement on which a judgment or decision can be based.

Critical control point (CCP)   A point, step, or procedure at which control can be applied and a food safety hazard can be prevented, eliminated, or reduced to acceptable levels.

Critical defect   A deviation at a CCP that may result in a hazard.

Critical limit   A criterion that must be met for each preventive measure associated with a critical control point.

Cross-contamination   The transfer of harmful microorganisms from one food to another by means of a nonfood surface, such as utensils, equipment, or human hands.

Danger zone   The temperature range between 40°F and 140°F (7.2°C and 60°C) within which most bacteria experience their best growth and reproduction.

---

*Adapted with permission from *Applied Foodservice Sanitation, Fourth Edition.* Copyright © 1992 by the Education Foundation of the National Restaurant Association.

Decline phase   The phase of bacteria growth, following the stationary phase, in which the rate of death within the colony exceeds the rate of reproduction and the number of living cells begins to decrease.

Deviation   Failure to meet a critical limit.

Fahrenheit   A temperature scale related to Celsius by the formula $(9/5 \times °Celsius) + 32° = °Fahrenheit$.

Foodborne illness   Disease or injury occurring as a result of consumption of contaminated food.

Food poisoning   A general term for intoxication or infection caused by consumption of contaminated food.

HACCP plan   The written document, based on the principles of HACCP, that delineates the procedures to be followed to assure the control of a specific process or procedure.

HACCP system   The result of the implementation of the HACCP plan.

HACCP team   The group of people who are responsible for developing and implementing an HACCP plan.

Hazard   A biological, chemical, or physical property that may cause a food to be unsafe for consumption.

Hygiene   Practices necessary for establishing and maintaining good health.

Incubation period   The phase in the course of an infection between the invasion of the host by the pathogen and the appearance of the symptoms of illness.

Infection   Disease caused by invasion of a host by living pathogenic organisms, which multiply within the body.

Intoxication   Disease caused by consumption of poisons, which may be chemical, naturally occurring in food, or produced by pathogenic microorganisms.

Lag phase   The period of bacterial growth following transfer to a new environment, when adaptation to new conditions takes place and there is little or no increase in the number of cells in the colony.

Log phase   The period of bacterial growth following the lag phase, when multiplication rate is constant and rapid and the number of cells in the colony increases exponentially.

Micro-   Prefix denoting small size.

Microbe   A general term for microscopic organisms, particularly pathogens.

Microorganisms   Forms of life that can be seen only with the aid of a microscope, including bacteria, viruses, yeasts, algae, and single-celled animals.

Monitor   To conduct a planned sequence of observations or measurements to assess whether a CCP is under control and to produce an accurate record for future use in verification.

Organism   An individual living thing.

Outbreak   The development of foodborne illness in two or more people who have eaten a common food that is shown by laboratory analysis to be the source of the illness. One case of botulism qualifies as an outbreak.

Personal hygiene   Individual cleanliness and habits that contribute to healthful conditions.

pH   A measure of the acidity or alkalinity of solutions: pH 7 is neutral, below 7 is acidic, and 7 to 14 is alkaline.

Potentially hazardous food (PHF) Any food that consists in whole or in part of milk or milk products, eggs, meat, poultry, fish, shellfish, edible crustacea, or other ingredients in a form capable of supporting rapid growth of infectious or toxicogenic microorganisms.

Preventive measure Physical, chemical, or other factors that can be used to control an identified health hazard.

Risk An estimate of the likely occurrence of a hazard.

Sanitary Free of disease-causing organisms and other harmful substances.

Sanitization The reduction of the number of pathogenic microorganisms on a surface to levels accepted as safe by regulatory authorities.

Stationary phase The period of bacterial growth, following the log phase, in which the number of bacterial cells remains more or less constant, as cells compete for space and nourishment.

Virus Any of a large group of infectious agents, lacking dependent metabolism and requiring a living host in order to reproduce, consisting of DNA or RNA in a protein shell.

Water activity Expression of amount of moisture available to aid bacterial growth.

Food Protection Quizzes

Food Protection Quiz 1
POTENTIALLY HAZARDOUS FOODS

1. Which of the following food items is potentially hazardous?
   a. Box of crackers
   b. Bottle of grapefruit juice
   c. Roast beef sandwich
   d. Cooked rice
   e. Both c and d

2. Which group contains only potentially hazardous foods?
   a. Cold cooked potatoes, warm cooked rice, refried beans, chicken salad
   b. Uncooked shrimp egg rolls, tofu, dry rice, egg drop soup, fresh garlic in oil
   c. Cooked spaghetti, vegetarian tomato sauce, dinner rolls, meatballs
   d. Hot beef stew, alfalfa sprouts, mashed potatoes, salad greens

3. Potentially hazardous foods are considered high risk because they:
   a. Spoil more quickly
   b. Allow bacteria to survive the cooking process
   c. Have properties that support rapid bacterial growth
   d. Are the kinds of food we like to eat

4. All of the following are considered to be potentially hazardous foods *except?*
   a. Apple pie
   b. Whole baked potato
   c. Well-done roast beef
   d. Turkey gravy

5. Disease-causing bacteria are always found on fresh, Grade A meat and poultry.
       True   False   (circle one)

6. Whole eggs with clean, intact shells are safe to be kept at room temperature.
       True   False   (circle one)

7. Re-serving a basket of nacho chips is considered OK because nacho chips are not considered potentially hazardous.
       True   False   (circle one)

297

8. Which of the following foods is not considered potentially hazardous?
   a. Partially cooked bacon
   b. Hard-boiled, air-dried eggs with shells intact
   c. Fresh garlic in oil
   d. Whipped butter

9. What types of food are most commonly implicated in foodborne illness?
   a. High acid foods such as tomatoes
   b. High protein foods such as poultry, beef, fish, pork, and dairy products
   c. Low acid foods such as mushrooms, tuna, and corn

ANSWER KEY

1.   e
2.   a
3.   c
4.   a
5.   T
6.   F
7.   F
8.   b
9.   b

Food Protection Quiz 2

CRITICAL CONTROL POINTS—EMPLOYEE HEALTH AND HYGIENE

1. In which of the following situations did the manager of a foodservice establishment act correctly?
   a. He sent home an employee who was sneezing and had a runny nose.
   b. He assigned a salad worker who had diarrhea to the dish washing room.
   c. He provided a clean, dry bandage to a sandwich maker who had an infected cut on her finger so that she could continue working.
   d. He restricted a cook who had an infected burn to the preparation of desserts.

2. The proper hand-washing procedure included all of these steps except:
   a. Use soap and warm running water while rubbing your hands vigorously for 20 seconds
   b. Wash all surfaces and rinse well under running water
   c. Dry your hands with a paper towel or electric dryer
   d. Turn off water with bare hands

3. A manager notices that a cook has a small, open cut on his finger and is preparing a tossed salad. The manager should
   a. Do nothing if the cut is showing signs of healing
   b. Instruct the cook to bandage the cut and wear plastic gloves
   c. Restrict the cook from preparation duties
   d. If the cut is not bleeding, cover it with a bandage

4. Eating and smoking are not allowed in the kitchen or while working because
   a. It looks bad in front of the customers
   b. Crumbs and ashes can drop intio the food
   c. It pollutes the air
   d. Touching the mouth contaminates the hands

5. It is estimated that 25 percent of all foodborne illness is caused by
   a. Improper label instructions
   b. Poor personal hygiene practices
   c. Not wearing the proper uniform
   d. Food stored under shelving

6. All foodservice employees should always wash their hands with soap and water
   a. Before starting work and after all breaks
   b. After sneezing, coughing, smoking, and touching hair
   c. After using the rest room
   d. After handling raw foods and soiled utensils
   e. All of the above

7. Which of the following is a substitute for proper, frequent hand washing?
   a. Instant hand sanitizer
   b. Disposable gloves, changed every hour
   c. Dipping hands into a bucket of bleach water
   d. Wiping hands on a dry cloth every time hands appear to be dirty
   e. a and c only
   f. None of the above

8. Which of the following is the proper use of plastic gloves?
   a. Wash the glove just as you wash your hands.
   b. Change gloves every hour.
   c. Change gloves under same circumstances you would wash hands.
   d. Wearing gloves is a substitute for hand washing.

9. An employee working night shift has been trying to unclog a sink with drain cleaner. A large party now enters the establishment. The cook needs help in the salad prep area. What should the employee do before assisting the cook?
   Rank 1 to 4 in order of sequence:
   Put on a clean apron
   Properly wash hands
   Place drain cleaner back in its proper location
   Begin salad prep

ANSWER KEY
1.   a
2.   d
3.   b
4.   d
5.   b
6.   e
7.   f
8.   c
9.   3, 2, 1, 4

Food Protection Quiz 3

CRITICAL ITEMS — CROSS-CONTAMINATION

1. Cross contamination occurs when
   a. Employees fail to follow proper hand-washing procedures
   b. Raw meats are stored below cooked products
   c. Sanitizing solutions are at proper concentration
   d. Separate cutting boards are used for cooked and raw products

2. Between uses on the serving line, utensils should be stored
   a. On a plate next to the food
   b. In the food with the handle extended out of the food
   c. In a container of hot water with all of the other utensils
   d. In a clean pan next to the serving station

3. At the grill station, there are many opportunities for cross-contamination between raw products and products that will not receive further cooking. What is the most practical way to prevent cross-contamination?
   a. Wear gloves while handling food
   b. Use separate utensils for each raw product and do not touch raw products with hands
   c. Use separate cutting boards for raw and cooked products
   d. Assign two people to the station: one to handle raw products, and one to handle cooked and prepared items

4. Which of the following is a potential source of cross-contamination?
   a. Cutting boards
   b. Slicer
   c. Improper storage of raw food products
   d. A thermometer used to check the internal temperature of food
   e. All of the above

5. Cross-contamination often occurs when different foods (particularly poultry products) are prepared on the same surface without sanitizing between each use.
   True   False   (circle one)

6. Which of the following statements presents the best reason for using separate cutting boards when preparing raw and cooked foods?
   a. Using separate cutting boards contributes to work simplification.
   b. The juices from cooked foods make wooden cutting boards slippery.
   c. Bacteria from uncooked foods may be transferred to cooked foods by means of a cutting board.
   d. Using separate cutting boards will prevent them from wearing out.

7. To prevent cross-contamination when handling glasses, silverware, and plates, you should always
   a. Grasp glasses from the base
   b. Pick up spoons by the handles
   c. Handle plates from the bottom
   d. All of the above

8. When tasting food it is OK to
    a.   Taste from the stirring spoon
    b.   Dip your finger in the food to taste
    c.   Use a clean metal or plastic spoon each time
    d.   Use the same metal tasting spoon, after rinsing under faucet

ANSWER KEY

1.   a
2.   b
3.   b
4.   e
5.   T
6.   c
7.   d
8.   c

Food Protection Quiz 4

CRITICAL ITEMS — TIME/TEMPERATURE

1. The two most critical limits over which the food manager has the most control are
   a. Time and temperature
   b. pH and water activity
   c. Time and pH
   d. Temperature and water activity

2. Potentially hazardous foods should be reheated to
   a. 140°F within 2 hours
   b. 140°F within 1 hour
   c. 165°F within 4 hours
   d. 165°F within 2 hours

3. Poultry and stuffed products, such as stuffed pork chops and stuffed pasta, are to be cooked to an internal temperature of
   a. 165°F
   b. 155°F
   c. 150°F
   d. 140°F

4. When reheating poultry products, you do not have to reheat to 165°F if the product was originally cooked to that temperature.
       True   False   (circle one)

5. Which of the following is the minimum temperature at which eggs and egg products need to be cooked?
   a. 155°F
   b. 145°F
   c. 140°F
   d. 130°F

6. The temperature danger zone for food is between
   a. 35°F and 135°F
   b. 40°F and 140°F
   c. 45°F and 145°F
   d. 45°F and 165°F

7. Pork and pork products are to be cooked to an internal temperature of
   a. 140°F
   b. 145°F
   c. 155°F
   d. 165°F

8. For cooking hamburgers, the internal temperature should reach
   a. 135°F
   b. 145°F
   d. 155°F
   e. 165°F

9. Of the following, which is the most acceptable method of cooking and hot holding rare prime rib or other beef roasts: cook to an internal temperature of
   a. 130°F and hold on the warm back section of the stove until ready to serve
   b. 130°F and hot hold under infrared heat lamps until ready to serve
   c. 130°F and hot hold at 130°F in a heated compartment for 121 minutes before serving
   d. 130°F and cover loosely with foil and hold until ready to serve

10. Potentially hazardous cold foods (such as macaroni salad, potato salad, French toast batter, and sliced meats) must be stored at ____°F or colder.

11. Potentially hazardous hot foods (such as green chili, gravy, meat loaf, beans, soups, and cooked pasta) must be stored at ____°F or hotter.

12. The minimum internal temperature required when reheating leftovers is ____°F.

ANSWER KEY

1.	a
2.	d
3.	a
4.	F
5.	c
6.	b
7.	c
8.	c
9.	c
10.	40
11.	140
12.	165

Food Protection Quiz 5

CRITICAL ITEMS—COOKING, COOLING, REHEATING, AND HOLDING

1. Using the HACCP principles, which of the following is the best way of judging whether or not a roasting chicken is done?
   a. By noting the color and appearance of the surface of the chicken
   b. By using a timer to record the length of the cooking process
   c. By using a thermometer inserted in the center of the chicken breast
   d. By cutting into the chicken between the leg and the breast to see if the juices are clear

2. When thawing and then cooking in the microwave, which of the following is the most acceptable?
   a. Refrigerate the food immediately after thawing.
   b. Cook immediately after thawing, add 25 degrees to the product's final cooking temperature.
   c. Rotate the food several times to thaw evenly, then hold the product in the refrigerator until it is ready to cook.
   d. Rotate the food several times to thaw evenly, then continue cooking to the product's normal cooking temperature. Let stand covered for two minutes for even temperature.

3. After cooking, potentially hazardous foods must be cooled to 70°F within 2 hours and then cooled to 40°F within a maximum of
   a. 1 hour
   b. 2 hours
   c. 4 hours
   d. 6 hours

4. Of the following methods of cooling thick, dense potentially hazardous foods, such as refried beans or rice, which is the *most* acceptable?
   a. Transfer the product to smaller pans, place pans in an ice bath, stir every 15 minutes, when the product is cooled to 40°F transfer to the refrigerator.
   b. Place the pot in a freezer and move it to a refrigerator after 3 hours.
   c. Transfer the product into 2-inch deep pans with product depth of 2 inches or less and place the pans in a high-velocity air flow refrigerator at 40°F.
   d. Transfer the hot product into several 2-inch shallow pans, cover, (to reduce chance of debris contaminating the food), and place in the walk-in.

5. Which of the following methods is *not* acceptable for thawing frozen foods?
   a. Gradual thawing under refrigeration
   b. Rapid thawing with hot water
   c. Gradual thawing under cold running water
   d. Gradual thawing at room temperature
   e. Both b and d

6. Which of the following is the best method to properly cool 5 gallons of freshly cooked green chili with pork?
   a. Let the chili sit on a table until the metal spoon is cool to the touch, then place in the walk-in, covered so that the heat from the food doesn't cause the ambient air temperature to rise.

b. Pour the chili immediately into a 5-gallon plastic container, cover and immediately place in the walk-in.

c. Pour the hot chili into several 2-inch shallow pans, cover (to reduce chance of debris contaminating the food) and place in the walk-in.

d. Immediately place pot of chili in an ice bath and surround with ice to the level of food, leave uncovered and stir every 15 minutes until temperature drops to 40°F, cover and place in the walk-in.

7. The following method should be used to reheat the green chili the next day.
   a. Place cold chili in a steam table pan in a steam table which has been set at 165°F, cover food and heat for at least 2 hours.
   b. Put chili in a warm steam table early in the day to allow chili to simmer slowly until lunch time.
   c. Rapidly reheat the chili on the stove top until it bubbles and steams, then place in preheated steam table.
   d. Reheat quickly on the stove top until it reaches 165°F all the way through, then pour into steam table pans in an accurate 140°F steam table.

8. At what temperature should the knob on the steam table be set to reheat leftover foods?
   a. 140°F
   b. 150°F
   c. 165°F
   d. Depends on the food
   e. None of the above

ANSWER KEY

1.	c
2.	b
3.	c
4.	c
5.	e
6.	d
7.	d
8.	e

Food Protection Quiz 6
CONTROL PROCEDURES — SANITIZING

1. When using a chlorine sanitizer solution in a three-compartment sink, what is the acceptable strength?
   a. 12.5–25 ppm
   b. 50–100 ppm
   c. 50–200 ppm
   d. 100–200 ppm

2. To sanitize a surface properly, the foodservice employee should
   a. Wash with warm soapy water, sanitize, rinse with clean water, air dry.
   b. Wash with warm soapy water, rinse with clean water, towel dry.
   c. Wash with warm soapy water, rinse with clean water, sanitize, air dry.
   d. By using a double-strength sanitizer, washing is not necessary.

3. When should cutting boards, meat slicers, knives, and other utensils be sanitized?
   a. During slow periods
   b. When the kitchen is opened each day
   c. After each use
   d. When you have prepared raw vegetables

4. The water temperature of a sanitizing solution should be
   a. 150°F to kill all harmful bacteria
   b. As warm as your hands can comfortably stand
   c. 75°F in order to prevent chlorine dissipation
   d. Outside the danger zone

5. When using wiping cloths to clean food contact surfaces, they should be
   a. Rinsed with hot water between uses
   b. Washed, rinsed, and soaked in a sanitizing solution
   c. Soaked in cold water between each use

6. What factor would limit the effectiveness of a chemical sanitizer?
   a. Hard water
   b. Water at a temperature of 120°F
   c. Detergent residue

7. To help in supervising the cleaning and making sure that foodservice workers follow proper cleaning and sanitizing procedures, the manager should implement a
   a. Point system for cleaning defects
   b. Code of conduct for cleaners
   c. Self-inspection program

ANSWER KEY

1.   b
2.   c
3.   c
4.   c
5.   b
6.   c
7.   c

Food Protection Quiz 7

ESTABLISHING CORRECTIVE ACTION

1. Your prep cook reports to you that she has diarrhea; you are short staffed. What is an incorrect response?
   a. Send her home.
   b. Let her put dry stock away and do paperwork.
   c. Let her continue to work, because you don't have anyone to replace her.
   d. Let her work in the dish room, handling only soiled dishes.

2. While frying chicken, you discover that the cook is using the same tongs for placing raw chicken in the fryer and panning the cooked chicken. What is the correct response?
   a. Discard the chicken.
   b. Do nothing, since the hot oil will kill the bacteria on the tongs.
   c. Put the chicken in the warmer, since the hot chicken will kill the bacteria.
   d. Put the chicken on a sheet pan and reheat in an oven to 165°F, and instruct the cook to use separate tongs for raw and cooked products.

3. Leftover rolls are returned to the kitchen from breadbaskets at the customers' tables. What is the correct response?
   a. Discard rolls; they are considered contaminated.
   b. Serve the rolls to customers, since they are not potentially hazardous foods.
   c. Use them for bread crumbs and croutons, since they will be reheated.
   d. Use them for bread pudding if they are properly dried first.

4. The manager discovers that the freezer containing meat is not working; the product temperature has reached 65°F. What is a proper corrective action?
   a. Cook the food immediately and refreeze the meat, because the bacteria will be killed.
   b. Refreeze the meat quickly in another freezer to control bacterial growth.
   c. Donate the food to charity.
   d. Discard the food, since possible bacterial toxins will not be destroyed by cooking.

5. A line server has reported to you that the food temperature on the steam table has dropped from 140°F to 130°F in the last hour. What is an appropriate corrective action?
   a. Discard the food, because it has dropped into the danger zone.
   b. Turn up the steam table to 150°F and check the food in one-half hour.
   c. Reheat the food to 165°F and adjust the steam table temperature to maintain 140°F or above.
   d. Continue serving the product, since two-thirds of the product has already been sold.

6. Leftover foods from a buffet should be
   a. Refrigerated promptly and rapidly reheated to 165°F before placing in the steamtable
   b. Discarded
   c. Refrigerated and reheated to 145°F and then placed in the steam table
   d. Rapidly cooled to 40°F and rapidly reheated to 165°F before placing in the steam table

7. If you have doubts about any food product, you should
   a. Throw it out.
   b. Serve it to your customers and hope they don't get sick.
   c. Donate it to charity.

8. You arrive in the morning to discover that the chili made the previous day has an internal temperature of 58°F. You should
    a. Quickly reheat the product to 165°F and serve it for lunch.
    b. Quickly reheat to 165°F, then rapidly cool in an ice bath to 40°F, refrigerate and then serve within two days.
    c. Quickly reheat to 165°F, rapidly cool in an ice bath to 40°F, and then freeze.
    d. Throw it out.

ANSWER KEY
1.   c
2.   d
3.   a
4.   d
5.   c
6.   b
7.   a
8.   d

Food Protection Quiz 8

FOODBORNE ILLNESS

1. Viral infections such as Norwalk and Hepatitis A are frequently caused by
   a. Foodhandlers who do not wash their hands before handling foods
   b. Dirty kitchen and facility
   c. Contaminated meats
   d. Time and temperature abuse
   e. Thawing frozen food at room temperature

2. Salmonella organisms are likely to be found in which of the following?
   a. Lettuce
   b. Raw chicken
   c. Cooked beans
   d. Chilled applesauce

3. The foodservice manager can best control bacterial spore formation by
   a. Cooking food to its proper temperature
   b. Rapidly cooling food to 40°F and reheating food to 165°F
   c. Storing food below 40°F
   d. Preventing cross-contamination

4. A key factor in the prevention of bacterial growth is
   a. The application of appropriate chemicals
   b. The breeding of less dangerous strains of microorganisms
   c. Time and temperature control of food

5. In a foodservice operation, the greatest risk to safe food is (are)
   a. The facilities
   b. The food itself
   c. The people, including employees and customers

6. The following is an example of a protection against cross-contamination
   a. Time and temperature control of food
   b. Separate cutting boards for raw and cooked products
   c. Ill employees restricted from handling food

7. A key factor in the control and prevention of foodborne illness is the individual food-handler who is
   a. Trained and motivated to follow safe practices
   b. Periodically examined by a public health officer
   c. Interviewed by the foodservice manager before employment
   d. Can pass a food safety test

8. The microorganisms that are of most concern to foodservice managers are
   a. Molds
   b. Viruses
   c. Bacteria

9. For rapid multiplication, all bacteria need
   a. Oxygen
   b. Living cells to reproduce
   c. Food, moisture, and favorable temperatures

10. Viruses differ from bacteria because viruses
   a. Need a living host to reproduce
   b. Need potentially hazardous food in which to grow
   c. Need very high temperatures to survive

ANSWER KEY

1.	a
2.	b
3.	b
4.	c
5.	c
6.	b
7.	a
8.	c
9.	c
10.	a

Index